Algorithms Illuminated
Part 4: Algorithms for NP-Hard Problems

Tim Roughgarden

First Edition

Cover image:
Norman Zammitt, *Blue Burning*, 1982
acrylic on canvas; 84 x 168 1/2 in. (213.36 x 427.99 cm)
San Francisco Museum of Modern Art, Gift of Judge and Mrs. William J. Lasarow
© Estate of Norman Zammitt
Photograph: Katherine Du Tiel

ISBN: 978-0-9992829-6-0 (Paperback)
ISBN: 978-0-9992829-7-7 (ebook)

Library of Congress Control Number: 2017914282

Soundlikeyourself Publishing, LLC
New York, NY
soundlikeyourselfpublishing@gmail.com
www.algorithmsilluminated.org

Contents

Preface

This book is the fourth in a series based on my online algorithms courses that have been running regularly since 2012, which in turn are based on an undergraduate course that I taught many times at Stanford University. *Part 4* assumes at least some familiarity with asymptotic analysis and big-O notation, graph search and shortest-path algorithms, greedy algorithms, and dynamic programming (all covered in *Parts 1–3*).

What We'll Cover in This Book

Algorithms Illuminated, Part 4 is all about NP-hard problems and what to do about them.

Algorithmic tools for tackling NP-hard problems. Many real-world problems are "NP-hard" and appear unsolvable by the types of always-correct and always-fast algorithms that have starred in the first three parts of this book series. When an NP-hard problem shows up in your own work, you must compromise on either correctness or speed. We'll see techniques old (like greedy algorithms) and new (like local search) for devising fast heuristic algorithms that are "approximately correct," with applications to scheduling, influence maximization in social networks, and the traveling salesman problem. We'll also cover techniques old (like dynamic programming) and new (like MIP and SAT solvers) for developing correct algorithms that improve dramatically on exhaustive search; applications here include the traveling salesman problem (again), finding signaling pathways in biological networks, and television station repacking in a recent and high-stakes spectrum auction in the United States.

Recognizing NP-hard problems. This book will also train you to quickly recognize an NP-hard problem so that you don't inadvertently waste time trying to design a too-good-to-be-true algorithm for it. You'll acquire familiarity with many famous and basic NP-hard problems, ranging from satisfiability to graph coloring to the Hamiltonian path problem. Through practice, you'll learn the tricks of the trade in proving problems NP-hard via reductions.

For a more detailed look into the book's contents, check out the "Upshot" sections that conclude each chapter and highlight the most important points. The "Field Guide to Algorithm Design" on page 236 provides a bird's-eye view of how the topics of this book fit into the bigger algorithmic picture.

The starred sections of the book are the most advanced ones. The time-constrained reader can skip these sections on a first reading without any loss of continuity.

Topics covered in the first three parts. *Algorithms Illuminated, Part 1* covers asymptotic notation (big-O notation and its close cousins), divide-and-conquer algorithms and the master method, randomized QuickSort and its analysis, and linear-time selection algorithms. *Part 2* is about data structures (heaps, balanced search trees, hash tables, bloom filters), graph primitives (breadth- and depth-first search, connectivity, shortest paths), and their applications (ranging from deduplication to social network analysis). *Part 3* focuses on greedy algorithms (scheduling, minimum spanning trees, clustering, Huffman codes) and dynamic programming (knapsack, sequence alignment, shortest paths, optimal search trees).

Skills You'll Learn From This Book Series

Mastering algorithms takes time and effort. Why bother?

Become a better programmer. You'll learn several blazingly fast subroutines for processing data as well as several useful data structures for organizing data that you can deploy directly in your own programs. Implementing and using these algorithms will stretch and improve your programming skills. You'll also learn general algorithm design paradigms that are relevant to many different problems across different domains, as well as tools for predicting the performance of such algorithms. These "algorithmic design patterns" can help you come up with new algorithms for problems that arise in your own work.

Sharpen your analytical skills. You'll get lots of practice describing and reasoning about algorithms. Through mathematical analysis, you'll gain a deep understanding of the specific algorithms and data structures that these books cover. You'll acquire facility with several mathematical techniques that are broadly useful for analyzing algorithms.

Think algorithmically. After you learn about algorithms, you'll start seeing them everywhere, whether you're riding an elevator, watching a flock of birds, managing your investment portfolio, or even watching an infant learn. Algorithmic thinking is increasingly useful and prevalent in disciplines outside of computer science, including biology, statistics, and economics.

Literacy with computer science's greatest hits. Studying algorithms can feel like watching a highlight reel of many of the greatest hits from the last sixty years of computer science. No longer will you feel excluded at that computer science cocktail party when someone cracks a joke about Dijkstra's algorithm. After reading these books, you'll know exactly what they mean.

Ace your technical interviews. Over the years, countless students have regaled me with stories about how mastering the concepts in these books enabled them to ace every technical interview question they were ever asked.

How These Books Are Different

This series of books has only one goal: *to teach the basics of algorithms in the most accessible way possible.* Think of them as a transcript of what an expert algorithms tutor would say to you over a series of one-on-one lessons.

There are a number of excellent more traditional and encyclopedic textbooks about algorithms, any of which usefully complement this book series with additional details, problems, and topics. I encourage you to explore and find your own favorites. There are also several books that, unlike these books, cater to programmers looking for ready-made algorithm implementations in a specific programming language. Many such implementations are freely available on the Web as well.

Who Are You?

The whole point of these books and the online courses upon which they are based is to be as widely and easily accessible as possible. People of all ages, backgrounds, and walks of life are well represented in my online courses, and there are large numbers of students (high-school, college, etc.), software engineers (both current and aspiring), scientists, and professionals hailing from all corners of the world.

This book is not an introduction to programming, and ideally you've acquired basic programming skills in a standard language (like Java, Python, C, Scala, Haskell, etc.). If you need to beef up your programming skills, there are several outstanding free online courses that teach basic programming.

We also use mathematical analysis as needed to understand how and why algorithms really work. The freely available book *Mathematics for Computer Science*, by Eric Lehman, F. Thomson Leighton, and Albert R. Meyer, is an excellent and entertaining refresher on mathematical notation (like $\sum$ and $\forall$), the basics of proofs (induction, contradiction, etc.), discrete probability, and much more.

Additional Resources

These books are based on online courses that are currently running on the Coursera and EdX platforms. I've made several resources available to help you replicate as much of the online course experience as you like.

Videos. If you're more in the mood to watch and listen than to read, check out the YouTube video playlists available at www.algorithmsilluminated.org. These videos cover all the topics in this book series, as well as additional advanced topics. I hope they exude a contagious enthusiasm for algorithms that, alas, is impossible to replicate fully on the printed page.

Quizzes. How can you know if you're truly absorbing the concepts in this book? Quizzes with solutions and explanations are scattered throughout the text; when you encounter one, I encourage you to pause and think about the answer before reading on.

End-of-chapter problems. At the end of each chapter, you'll find several relatively straightforward questions that test your understanding, followed by harder and more open-ended challenge problems. Hints or solutions to all of these problems (as indicated by an "*(H)*" or "*(S)*," respectively) are included at the end of the book. Readers can interact with me and each other about the end-of-chapter problems through the book's discussion forum (see below).

Programming problems. Several of the chapters conclude with suggested programming projects whose goal is to help you develop a detailed understanding of an algorithm by creating your own working implementation of it. Data sets, along with test cases and their solutions, can be found at `www.algorithmsilluminated.org`.

Discussion forums. A big reason for the success of online courses is the opportunities they provide for participants to help each other understand the course material and debug programs through discussion forums. Readers of these books have the same opportunity via the forums available at `www.algorithmsilluminated.org`.

Acknowledgments

These books would not exist without the passion and hunger supplied by the hundreds of thousands of participants in my algorithms courses over the years. I am particularly grateful to those who supplied detailed feedback on an earlier draft of this book: Tonya Blust, Yuan Cao, Leslie Damon, Tyler Dae Devlin, Roman Gafiteanu, Blanca Huergo, Jim Humelsine, Tim Kearns, Vladimir Kokshenev, Bayram Kuliyev, Clayton Wong, Lexin Ye, and Daniel Zingaro. Thanks also to several experts who provided technical advice: Amir Abboud, Vincent Conitzer, Christian Kroer, Aviad Rubinstein, and Ilya Segal.

I always appreciate suggestions and corrections from readers. These are best communicated through the discussion forums mentioned above.

Tim Roughgarden
New York, NY
June 2020

Chapter 19

What Is NP-Hardness?

Introductory books on algorithms, including *Parts 1–3* of this series, suffer from selection bias. They focus on computational problems that are solvable by clever, fast algorithms—after all, what's more fun and empowering to learn than an ingenious algorithmic short-cut? The good news is that many fundamental and practically relevant problems fall into this category: sorting, graph search, shortest paths, Huffman codes, minimum spanning trees, sequence alignment, and so on. But it would be fraudulent to teach you only this cherry-picked collection of problems while ignoring the spectre of computational intractability that haunts the serious algorithm designer or programmer. Sadly, there are many important computational problems, including ones likely to show up in your own projects, for which no fast algorithms are known. Even worse, we can't expect any future algorithmic breakthroughs for these problems, as they are widely believed to be intrinsically difficult and unsolvable by any fast algorithm.

Newly aware of this stark reality, two questions immediately come to mind. First, how can you recognize such hard problems when they appear in your own work, so that you can adjust your expectations accordingly and avoid wasting time looking for a too-good-to-be-true algorithm? Second, when such a problem is important to your application, how should you revise your ambitions, and what algorithmic tools can you apply to achieve them? This book will equip you with thorough answers to both questions.

19.1 MST vs. TSP: An Algorithmic Mystery

Hard computational problems can look a lot like easy ones, and telling them apart requires a trained eye. To set the stage, let's rendezvous with a familiar friend (the minimum spanning tree problem) and meet its more demanding cousin (the traveling salesman problem).

19.1.1   The Minimum Spanning Tree Problem

One famous computational problem solvable by a blazingly fast algorithm is the *minimum spanning tree (MST)* problem (covered in Chapter 15 of *Part 3*).[1]

Problem: Minimum Spanning Tree (MST)

Input: A connected undirected graph $G = (V, E)$ and a real-valued cost c_e for each edge $e \in E$.

Output: A spanning tree $T \subseteq E$ of G with the minimum-possible sum $\sum_{e \in T} c_e$ of edge costs.

Recall that a graph $G = (V, E)$ is *connected* if, for every pair $v, w \in V$ of vertices, the graph contains a path from v to w. A *spanning tree* of G is a subset $T \subseteq E$ of edges such that the subgraph (V, T) is both connected and acyclic. For example, in the graph

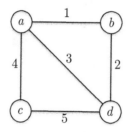

the minimum spanning tree comprises the edges (a, b), (b, d), and (a, c), for an overall cost of 7.

A graph can have an exponential number of spanning trees, so exhaustive search is out of the question for all but the smallest graphs.[2] But the MST problem *can* be solved by clever fast algorithms,

---

[1]To review, a *graph* $G = (V, E)$ has two ingredients: a set V of *vertices* and a set E of *edges*. In an *undirected* graph, each edge $e \in E$ corresponds to an unordered pair $\{v, w\}$ of vertices (written as $e = (v, w)$ or $e = (w, v)$). In a *directed* graph, each edge (v, w) is an ordered pair, with the edge directed from v to w. The numbers $|V|$ and $|E|$ of vertices and edges are usually denoted by n and m, respectively.

[2]For example, *Cayley's formula* is a famous result from combinatorics stating that the n-vertex complete graph (in which all the $\binom{n}{2}$ possible edges are present) has exactly n^{n-2} different spanning trees. This is bigger than the estimated number of atoms in the known universe when $n \geq 50$.

such as Prim's and Kruskal's algorithms. Deploying appropriate data structures (heaps and union-find, respectively), both algorithms have blazingly fast implementations, with a running time of $O((m + n)\log n)$, where m and n are the number of edges and vertices of the input graph, respectively.

19.1.2   The Traveling Salesman Problem

Another famous problem, absent from *Parts 1–3* but prominent in this book, is the *traveling salesman problem (TSP)*. Its definition is almost the same as that of the MST problem, except with *tours—* simple cycles that span all vertices—playing the role of spanning trees.

Problem: Traveling Salesman Problem (TSP)

Input: A complete undirected graph $G = (V, E)$ and a real-valued cost c_e for each edge $e \in E$.[3]

Output: A tour $T \subseteq E$ of G with the minimum-possible sum $\sum_{e \in T} c_e$ of edge costs.

Formally, a *tour* is a cycle that visits every vertex exactly once (with two edges incident to each vertex).

Quiz 19.1

In an instance $G = (V, E)$ of the TSP with $n \geq 3$ vertices, how many distinct tours $T \subseteq E$ are there? (In the answers below, $n! = n \cdot (n-1) \cdot (n-2) \cdots 2 \cdot 1$ denotes the factorial function.)

a) 2^n

b) $\frac{1}{2}(n-1)!$

---

[3]In a *complete* graph, all $\binom{n}{2}$ possible edges are present. The assumption that the graph is complete is without loss of generality, as an arbitrary input graph can be harmlessly turned into a complete graph by adding in all the missing edges and giving them very high costs.

c) $(n-1)!$

d) $n!$

(See Section 19.1.4 for the solution and discussion.)

If all else fails, the TSP can be solved by exhaustively enumerating all of the (finitely many) tours and remembering the best one. Try exhaustive search out on a small example.

Quiz 19.2

What is the minimum sum of edge costs of a tour of the following graph? (Each edge is labeled with its cost.)

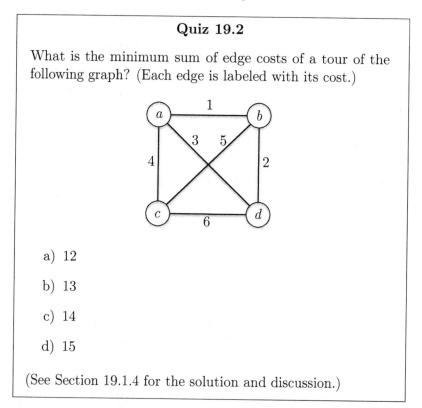

a) 12

b) 13

c) 14

d) 15

(See Section 19.1.4 for the solution and discussion.)

The TSP can be feasibly solved by exhaustive search for only the smallest of instances. *Can we do better?* Could there be, analogous to the MST problem, an algorithm that magically homes in on the minimum-cost needle in the exponential-size haystack of traveling salesman tours? Despite the superficial similarity of the statements of the two problems, the TSP appears to be far more difficult to solve than the MST problem.

19.1.3   Trying and Failing to Solve the TSP

I could tell you a cheesy story about, um, a traveling salesman, but this would do a disservice to the TSP, which is actually quite fundamental. Whenever you have a bunch of tasks to complete in a sequence, with the cost or time for carrying out a task dependent on the preceding task, you're talking about the TSP in disguise.

For example, tasks could represent cars to be assembled in a factory, with the time required to assemble a car equal to a fixed cost (for assembly) plus a setup cost that depends on how different the factory configurations are for this and the previous car. Assembling all the cars as quickly as possible boils down to minimizing the sum of the setup costs, which is exactly the TSP.

For a very different application, imagine that you've collected a bunch of overlapping fragments of a genome and would like to reverse engineer their most plausible ordering. Given a "plausibility measure" that assigns a cost to each fragment pair (for example, derived from the length of their longest common substring), this ordering problem also boils down to the TSP.[4]

Seduced by the practical applications and aesthetic appeal of the TSP, many of the greatest minds in optimization have, since at least the early 1950s, devoted a tremendous amount of effort and computation to solving large-scale instances of the TSP.[5] Despite the decades and intellectual firepower involved:

Fact

As of this writing (in 2020), there is no known fast algorithm for the TSP.

What do we mean by a "fast" algorithm? Back in *Part 1*, we agreed that:

---

[4]Both applications are arguably better modeled as traveling salesman *path* problems, in which the goal is to compute a minimum-cost cycle-free path that visits every vertex (without going back to the starting vertex). Any algorithm solving the TSP can be easily converted into one solving the path version of the problem, and vice versa (Problem 19.7).

[5]Readers curious about the history or additional applications of the TSP should check out the first four chapters of the book *The Traveling Salesman Problem: A Computational Study*, by David L. Applegate, Robert E. Bixby, Vašek Chvátal, and William J. Cook (Princeton University Press, 2006).

A "fast algorithm" is an algorithm whose worst-case running time grows slowly with the input size.

And what do we mean by "grows slowly"? For much of this book series, the holy grail has been algorithms that run in linear or almost-linear time. Forget about such blazingly fast algorithms—for the TSP, no one even knows of an algorithm that always runs in $O(n^{100})$ time on n-vertex instances, or even $O(n^{10000})$ time.

There are two competing explanations for the dismal state-of-the-art: (i) there is a fast algorithm for the TSP but no one's been smart enough to find it yet; or (ii) no such algorithm exists. We do not know which explanation is correct, though most experts believe in the second one.

Speculation

No fast algorithm for the TSP exists.

As early as 1967, Jack Edmonds wrote:

> I conjecture that there is no good algorithm for the traveling saleman [sic] problem. My reasons are the same as for any mathematical conjecture: (1) It is a legitimate mathematical possibility, and (2) I do not know.[6]

Unfortunately, the curse of intractability is not confined to the TSP. We'll see that many other practically relevant problems are similarly afflicted.

19.1.4   Solutions to Quizzes 19.1–19.2

Solution to Quiz 19.1

Correct answer: (b). There is an intuitive correspondence between vertex orderings (of which there are $n!$) and tours (which visit the vertices once each, in some order), so answer (d) is a natural guess. However, this correspondence counts each tour in $2n$ different ways:

---

[6]From the paper "Optimum Branchings," by Jack Edmonds (*Journal of Research of the National Bureau of Standards, Series B*, 1967). By a "good" algorithm, Edmonds means an algorithm with a running time bounded above by some polynomial function of the input size.

once for each of the n choices of the initial vertex and once for each of the two directions of traversing the tour. Thus, the total number of tours is $n!/2n = \frac{1}{2}(n-1)!$. For example, with $n = 4$, there are three distinct tours:

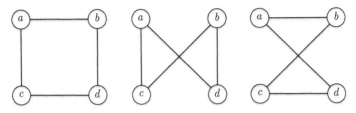

Solution to Quiz 19.2

Correct answer: (b). We can enumerate tours by starting with the vertex a and trying all six possible orderings of the other three vertices, with the understanding that the tour finishes by traveling from the last vertex back to a. (Actually, this enumeration counts each tour twice, once in each direction.) The results:

Vertex Ordering	Cost of Corresponding Tour
a, b, c, d or a, d, c, b	15
a, b, d, c or a, c, d, b	13
a, c, b, d or a, d, b, c	14

The shortest tour is the second one, with a total cost of 13.

19.2   Possible Levels of Expertise

Some computational problems are easier than others. The point of the theory of NP-hardness is to classify, in a precise sense, problems as either "computationally easy" (like the MST problem) or "computationally difficult" (like the TSP). This book is aimed both at readers looking for a white-belt primer on the topic and at those pursuing black-belt expertise. This section offers guidance on how to approach the rest of the book, as a function of your goals and constraints.

What are your current and desired levels of expertise in recognizing and tackling NP-hard problems?[7]

---

[7]What's up with the term "NP"? See Section 19.6.

(Level 0:) "What's an NP-hard problem?"

Level 0 is total ignorance—you've never heard of NP-hardness and are unaware that many practically relevant computational problems are widely believed to be unsolvable by any fast algorithm. If I've done my job, this book should be accessible even to level-0 readers.

(Level 1:) "Oh, the problem is NP-hard? I guess we should either reformulate the problem, scale down our ambitions, or invest a lot more resources into solving it."

Level 1 represents cocktail-party-level awareness and at least an informal understanding of what NP-hardness means.[8] For example, are you managing a software project with an algorithmic or optimization component? If so, you should acquire at least level-1 knowledge, in case one of your team members bumps into an NP-hard problem and wants to discuss the possible next steps. To raise your level to 1, study Sections 19.3, 19.4, and 19.6.

(Level 2:) "Oh, the problem is NP-hard? Give me a chance to apply my algorithmic expertise and see how far I can get."

The biggest marginal empowerment for software engineers comes from reaching level 2, and acquiring a rich toolbox for developing practically useful algorithms for solving or approximating NP-hard problems. Serious programmers should shoot for this level (or above). Happily, all the algorithmic paradigms that we developed for polynomial-time solvable problems in *Parts 1–3* are also useful for making headway on NP-hard problems. The goal of Chapters 20 and 21 is to bring you up to level 2; see also Section 19.4 for an overview and Chapter 24 for a detailed case study of the level-2 toolbox in action in a high-stakes application.

(Level 3:) "Tell me about your computational problem. [... listens carefully ...] My condolences, your problem is NP-hard."

At level 3, you can quickly recognize NP-hard problems when they arise in practice (at which point you can switch to applying your level-2 skills). You know several famous NP-hard problems and also

---

[8]Speaking, as always, about sufficiently nerdy cocktail parties!

how to prove that additional problems are NP-hard. Specialists in algorithms should master these skills. For example, I frequently draw on level-3 knowledge when advising colleagues, students, or engineers in industry on algorithmic problems. Chapter 22 provides a boot camp for upping your game to level 3; see also Section 19.5 for an overview.

(Level 4:) "Allow me to explain the P $\neq$ NP conjecture to you on this whiteboard."

Level 4, the most advanced level, is for budding theoreticians and anyone seeking a rigorous mathematical understanding of NP-hardness and the P vs. NP question. If that qualifier doesn't scare you off, the optional Chapter 23 is for you.

19.3 Easy and Hard Problems

An oversimplification of the "easy vs. hard" dichotomy proposed by the theory of NP-hardness is:

easy $\leftrightarrow$ can be solved with a polynomial-time algorithm;
hard $\leftrightarrow$ requires exponential time in the worst case.

This summary of NP-hardness overlooks several important subtleties (see Section 19.3.9). But ten years from now, if you remember only a few words about the meaning of NP-hardness, these are good ones.

19.3.1 Polynomial-Time Algorithms

To segue into the definition of an "easy" problem, let's recap the running times of some famous algorithms that you may have seen (for example, in *Parts 1–3*):

Problem	Algorithm	Running Time
Sorting	`MergeSort`	$O(n \log n)$
Strong Components	`Kosaraju`	$O(m + n)$
Shortest Paths	`Dijkstra`	$O((m + n) \log n)$
MST	`Kruskal`	$O((m + n) \log n)$
Sequence Alignment	`NW`	$O(mn)$
All-Pairs Shortest Paths	`Floyd-Warshall`	$O(n^3)$

The exact meaning of n and m is problem-specific, but in all cases they are closely related to the input size.[9] The key takeaway from this table is that, while the running times of these algorithms vary, *all of them are bounded above by some polynomial function of the input size.* In general:

Polynomial-Time Algorithms

A *polynomial-time algorithm* is an algorithm with worst-case running time $O(n^d)$, where n denotes the input size and d is a constant (independent of n).

The six algorithms listed at the beginning of this section are all polynomial-time algorithms (with reasonably small exponents d).[10] Do all natural algorithms run in polynomial time? No. For example, for many problems, exhaustive search runs in time exponential in the input size (as noted in footnote 2 for the MST problem). There's something special about the clever polynomial-time algorithms that we've studied so far.

19.3.2   Polynomial vs. Exponential Time

Don't forget that any exponential function eventually grows much faster than any polynomial function. There's a huge difference between typical polynomial and exponential running times, even for very small instances. The plot at the top of the next page (of the polynomial function $100n^2$ versus the exponential function 2^n) is representative.

Moore's law asserts that the computing power available for a given price doubles every 1–2 years. Does this mean that the difference between polynomial-time and exponential-time algorithms will disappear over time? Actually, the exact opposite is true! Our computational ambitions grow with our computational power, and as time goes on we consider increasingly large input sizes and suffer an increasingly big gulf between polynomial and exponential running times.

---

[9]In sorting, n denotes the length of the input array; in the four graph problems, n and m denote the number of vertices and edges, respectively; and in the sequence alignment problem, n and m denote the lengths of the two input strings.

[10]Remember that a logarithmic factor can be bounded above (sloppily) by a linear factor; for example, if $T(n) = O(n \log n)$, then $T(n) = O(n^2)$ as well.

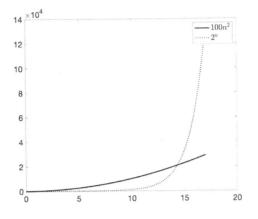

Imagine that you have a fixed time budget, like an hour or a day. How does the solvable input size scale with additional computing power? With a polynomial-time algorithm, it increases by a constant factor (such as from 1,000,000 to 1,414,213) with every doubling of your computing power.[11] With an algorithm that runs in time proportional to 2^n, where n is the input size, each doubling of computing power increases the solvable input size by only one (such as from 1,000,000 to 1,000,001)!

19.3.3   Easy Problems

The theory of NP-hardness defines "easy" problems as those solvable by a polynomial-time algorithm, or equivalently by an algorithm for which the solvable input size (for a fixed time budget) scales multiplicatively with increasing computational power:[12]

Polynomial-Time Solvable Problems

A computational problem is *polynomial-time solvable* if there is a polynomial-time algorithm that solves it correctly for every input.

---

[11]With a linear-time algorithm, you could solve problems that are twice as big; with a quadratic-time algorithm, $\sqrt{2} \approx 1.414$ times as big; with a cubic-time algorithm, $\sqrt[3]{2} \approx 1.26$ as big; and so on.

[12]This definition was proposed independently by Alan Cobham and Jack Edmonds (see footnote 6) in the mid-1960s.

For example, the six problems listed at the beginning of this section are all polynomial-time solvable.

Technically, a (useless-in-practice) algorithm that runs in $O(n^{100})$ time on size-n inputs counts as a polynomial-time algorithm, and a problem solved by such an algorithm qualifies as polynomial-time solvable. Turning this statement around, if a problem like the TSP is *not* polynomial-time solvable, there is not even an $O(n^{100})$-time or $O(n^{10000})$-time algorithm that solves it (!).

Courage, Definitions, and Edge Cases

The identification of "easy" with "polynomial-time solvable" is imperfect; a problem might be solved in theory (by an algorithm that technically runs in polynomial time) but not in reality (by an empirically fast algorithm), or vice versa. Anyone with the guts to write down a precise mathematical definition (like polynomial-time solvability) to express a messy real-world concept (like "easy to solve via computer in the physical world") must be ready for friction between the binary nature of the definition and the fuzziness of reality. The definition will inevitably include or exclude some edge cases that you wish had gone the other way, but this is no excuse to ignore or dismiss a good definition. Polynomial-time solvability has been unreasonably effective at classifying problems as "easy" or "hard" in a way that accords with empirical experience. With a half-century of evidence behind us, we can confidently say that natural polynomial-time solvable problems typically can be solved with practical general-purpose algorithms, and that problems believed to not be polynomial-time solvable typically require significantly more work and domain expertise.

19.3.4 Relative Intractability

Suppose we suspected that a problem like the TSP is "not easy," meaning unsolvable by any polynomial-time algorithm (no matter how large the polynomial). How would we amass evidence that this is, in fact, the case? The most convincing argument, of course, would be an airtight mathematical proof. But the status of the TSP remains in limbo to this day: No one has found a polynomial-time algorithm that solves it, nor has anyone found a proof that no such algorithm exists.

How can we develop a theory that usefully differentiates "tractable" and "intractable" problems despite our deficient understanding of what algorithms can do? The brilliant conceit behind the theory of NP-hardness is to classify problems based on their *relative* (rather than absolute) difficulty and to declare a problem as "hard" if it is "at least as hard as" an overwhelming number of other unsolved problems.

19.3.5   Hard Problems

The many failed attempts at solving the TSP (Section 19.1.3) provide circumstantial evidence that the problem may not be polynomial-time solvable.

Weak Evidence of Hardness

A polynomial-time algorithm for the TSP would solve a problem that has resisted the efforts of hundreds (if not thousands) of brilliant minds over many decades.

Can we better, meaning build a more compelling case of intractability? This is where the magic and power of NP-hardness comes in. The big idea is to show that a problem like the TSP is at least as hard as a vast array of unsolved problems from many different scientific fields—in fact, all problems for which you quickly know a solution when you see one. Such an argument would imply that a hypothetical polynomial-time algorithm for the TSP would automatically solve all these other unsolved problems, as well!

Strong Evidence of Hardness

A polynomial-time algorithm for the TSP would solve *thousands* of problems that have resisted the efforts of *tens (if not hundreds) of thousands* of brilliant minds over many decades.

In effect, the theory of NP-hardness shows that thousands of computational problems (including the TSP) are variations of the same problem in disguise, all destined to suffer identical computational fates. If you're trying to devise a polynomial-time algorithm for an NP-hard problem like the TSP, you're inadvertently attempting to also come up with such algorithms for these thousands of related problems.[13]

We call a problem *NP-hard* if there is strong evidence of intractability in the sense above:

NP-Hardness (Main Idea)

A problem is *NP-hard* if it is at least as difficult as every problem with easily recognized solutions.

This idea will be made 100% precise in Section 23.3.4; until then, we'll work with a provisional definition of NP-hardness that is phrased in terms of a famous mathematical conjecture, the "P $\neq$ NP conjecture."

19.3.6 The P $\neq$ NP Conjecture

Perhaps you've heard of the P $\neq$ NP conjecture. What is it, exactly? Section 23.4 provides the precise mathematical statement; for now, we'll settle for an informal version that should resonate with anyone who's had to grade student homework:

The P $\neq$ NP Conjecture (Informal Version)

Checking an alleged solution to a problem can be fundamentally easier than coming up with your own solution from scratch.

---

[13]Playing devil's advocate, hundreds (if not thousands) of brilliant minds have likewise failed to prove the other direction, that the TSP is *not* polynomial-time solvable. Symmetrically, doesn't this suggest that perhaps no such proof exists? The difference is that we seem far better at proving solvability (with fast algorithms known for countless problems) than unsolvability. Thus, if the TSP were polynomial-time solvable, it would be odd that we haven't yet found a polynomial-time algorithm for it; if not, no surprise that we haven't yet figured out how to prove it.

The "P" and "NP" in the conjecture refer to problems that can be solved from scratch in polynomial time and those whose solutions can be checked in polynomial time, respectively; for formal definitions, see Chapter 23.

For example, checking someone's proposed solution to a Sudoku or KenKen puzzle sure seems easier than working it out yourself. Or, in the context of the TSP, it's easy to verify that someone's proposed traveling salesman tour is good (with a total cost of, say, at most 1000) by adding up the costs of its edges; it's not so clear how you would quickly come up with your own such tour from scratch. Thus, intuition strongly suggests that the P $\neq$ NP conjecture is true.[14,15]

19.3.7   Provisional Definition of NP-Hardness

Provisionally, we'll call a problem NP-hard if, assuming that the P $\neq$ NP conjecture *is* true, it cannot be solved by any polynomial-time algorithm.

NP-Hard Problem (Provisional Definition)

A computational problem is *NP-hard* if a polynomial-time algorithm solving it would refute the P $\neq$ NP conjecture.

Thus, any polynomial-time algorithm for any NP-hard problem (such as the TSP) would automatically imply that the P $\neq$ NP conjecture is false and trigger an algorithmic bounty that seems too good to be true: a polynomial-time algorithm for every single problem for which solutions can be recognized in polynomial time. In the likely event that the P $\neq$ NP conjecture is true, no NP-hard problem is polynomial-time solvable, not even with an algorithm that runs in $O(n^{100})$ or $O(n^{10000})$ time on size-n inputs.

---

[14]We'll see in Problem 23.2 that the P $\neq$ NP conjecture is equivalent to Edmonds's conjecture (page 6) stating that the TSP cannot be solved in polynomial time.

[15]Why isn't it "obvious" that the P $\neq$ NP conjecture is true? Because the space of polynomial-time algorithms is unfathomably rich, with many ingenious inhabitants. (Perhaps you've come across Strassen's mind-blowing subcubic matrix multiplication algorithm, for example in Chapter 3 of *Part 1*?) Proving that none of the infinitely many candidate algorithms solve the TSP seems pretty intimidating!

19.3.8   Randomized and Quantum Algorithms

Our definition of polynomial-time solvability on page 11 contemplates only deterministic algorithms. As we know, randomization can be a powerful tool in algorithm design (for example, in the QuickSort algorithm). Can randomized algorithms escape the binds of NP-hardness?

More generally, what about much-hyped quantum algorithms? (As it turns out, randomized algorithms can be viewed as a special case of quantum algorithms.) It's true that large-scale, general-purpose quantum computers (if realized) would be a game-changer for a handful of problems, including the extremely important problem of factoring large integers. However, the factoring problem is not known or believed to be NP-hard, and experts conjecture that even quantum computers cannot solve NP-hard problems in polynomial time. The challenges posed by NP-hardness are not going away anytime soon.[16]

19.3.9   Subtleties

The oversimplified discussion at the beginning of this section (page 9) suggested that a "hard" problem would require exponential time to solve in the worst case. Our provisional definition in Section 19.3.7 says something different: An NP-hard problem is one that, assuming the P $\neq$ NP conjecture, cannot be solved by any polynomial-time algorithm.

The first discrepancy between the two definitions is that NP-hardness rules out polynomial-time solvability only if the P $\neq$ NP conjecture is true (and this remains an open question). If the conjecture is false, almost all the NP-hard problems discussed in this book are, in fact, polynomial-time solvable.

The second discrepancy is that, even in the likely event that the P $\neq$ NP conjecture is true, NP-hardness implies only that super-

---

[16]A majority of experts believe that every polynomial-time randomized algorithm can be *derandomized* and turned into an equivalent polynomial-time deterministic algorithm (perhaps with a larger polynomial in the running time bound). If true, the P $\neq$ NP conjecture would automatically apply to randomized algorithms as well.

By contrast, a majority of experts believe that quantum algorithms *are* fundamentally more powerful than classical algorithms (but not powerful enough to solve NP-hard problems in polynomial time). Isn't it amazing—and exciting—how much we still don't know?

polynomial (as opposed to exponential) time is required in the worst case to solve the problem.[17] However, for most natural NP-hard problems, including all those studied in this book, experts generally believe that exponential time is indeed required in the worst case. This belief is formalized by the "Exponential Time Hypothesis," a stronger form of the P $\neq$ NP conjecture (see Section 23.5).[18]

Finally, while 99% of the problems that you'll come across will be either "easy" (polynomial-time solvable) or "hard" (NP-hard), a few rare examples appear to lie in between. Thus, our "dichotomy" between easy and hard problems covers most, but not all, practically relevant computational problems.[19]

19.4   Algorithmic Strategies for NP-Hard Problems

Suppose you've identified a computational problem on which the success of your project rests. Perhaps you've spent the last several weeks throwing the kitchen sink at it—all the algorithm design paradigms you know, every data structure in the book, all the for-free primitives—but nothing works. Finally, you realize that the issue is not a deficiency of ingenuity on your part, it's the fact that the problem is NP-hard. Now you have an explanation of why your weeks of effort have come to naught, but that doesn't diminish the problem's significance to your project. What should you do?

19.4.1   General, Correct, Fast (Pick Two)

The bad news is that NP-hard problems are ubiquitous; right now, one might well be lurking in your latest project. The good news is that NP-hardness is not a death sentence. NP-hard problems can often

---

[17]Examples of running time bounds that are super-polynomial but subexponential in the input size n include $n^{\log_2 n}$ and $2^{\sqrt{n}}$.

[18]None of the computational problems studied in this book series require more than exponential time to solve, but other problems do. One famous example is the "halting problem," which can't be solved in any finite (let alone exponential) amount of time; see also Section 23.1.2.

[19]Two important problems that are believed to be neither polynomial-time solvable nor NP-hard are factoring (finding a non-trivial factor of an integer or determining that none exist) and the graph isomorphism problem (determining whether two graphs are identical up to a renaming of the vertices). Subexponential-time (but not polynomial-time) algorithms are known for both problems.

(but not always) be solved in practice, at least approximately, through sufficient investment of resources and algorithmic sophistication.

NP-hardness throws down the gauntlet to the algorithm designer and tells you where to set your expectations. You should not expect a general-purpose and always-fast algorithm for an NP-hard problem, akin to those we've seen for problems such as sorting, shortest paths, or sequence alignment. Unless you're lucky enough to face only unusually small or well-structured inputs, you're going to have to work pretty hard to solve the problem, and possibly also make some compromises.

What kinds of compromises? NP-hardness rules out algorithms with the following three desirable properties (assuming the $P \neq NP$ conjecture):

Three Properties (You Can't Have Them All)

1. *General-purpose.* The algorithm accommodates all possible inputs of the computational problem.

2. *Correct.* For every input, the algorithm correctly solves the problem.

3. *Fast.* For every input, the algorithm runs in polynomial time.

Accordingly, you can choose from among three types of compromises: compromising on generality, compromising on correctness, and compromising on speed. All three strategies are useful and common in practice.

The rest of this section elaborates on these three algorithmic strategies; Chapters 20 and 21 are deep dives into the latter two. As always, our focus is on powerful and flexible algorithm design principles that apply to a wide range of problems. You should take these principles as a starting point and run with them, guided by whatever domain expertise you have for the specific problem that you need to solve.

19.4.2   Compromising on Generality

One strategy for making progress on an NP-hard problem is to give up on general-purpose algorithms and focus instead on special cases of the problem relevant to your application. In the best-case scenario, you can identify domain-specific constraints on inputs and design an algorithm that is always correct and always fast on this subset of inputs. Graduates of the dynamic programming boot camp in *Part 3* have already seen two examples of this strategy.

Weighted independent set. In this problem, the input is an undirected graph $G = (V, E)$ and a nonnegative weight w_v for each vertex $v \in V$; the goal is to compute an independent set $S \subseteq V$ with the maximum-possible sum $\sum_{v \in S} w_v$ of vertex weights, where an *independent set* is a subset $S \subseteq V$ of mutually non-adjacent vertices (with $(v, w) \notin E$ for every $v, w \in S$). For example, if edges represent conflicts (between people, courses, etc.), independent sets correspond to conflict-free subsets. This problem is NP-hard in general, as we'll see in Section 22.5. The special case of the problem in which G is a path graph (with vertices $v_1, v_2, \ldots, v_n$ and edges $(v_1, v_2), (v_2, v_3), \ldots, (v_{n-1}, v_n)$) can be solved in linear time using a dynamic programming algorithm. This algorithm can be extended to accommodate all acyclic graphs (see Problem 16.3 of *Part 3*).

Knapsack. In this problem, the input is specified by $2n + 1$ positive integers: n item values $v_1, v_2, \ldots, v_n$, n item sizes $s_1, s_2, \ldots, s_n$, and a knapsack capacity C. The goal is to compute a subset $S \subseteq \{1, 2, \ldots, n\}$ of items with the maximum-possible sum $\sum_{i \in S} v_i$ of values, subject to having total size $\sum_{i \in S} s_i$ at most C. In other words, the objective is to make use of a scarce resource in the most valuable way possible.[20] This problem is NP-hard, as we'll see in Section 22.8 and Problem 22.7. There is an $O(nC)$-time dynamic programming algorithm for the problem; this is a polynomial-time algorithm in the special case in which C is bounded by a polynomial function of n.

> ### A Polynomial-Time Algorithm for Knapsack?
>
> Why doesn't the $O(nC)$-time algorithm for the knapsack problem refute the P $\neq$ NP conjecture? Because this is not a polynomial-time algorithm. The input

---

[20] For example, on which goods and services should you spend your paycheck to get the most value? Or, given an operating budget and a set of job candidates with differing productivity levels and requested salaries, whom should you hire?

size—the number of keystrokes needed to specify the input to a computer—scales with the number of *digits* in a number, not the *magnitude* of a number. It doesn't take a million keystrokes to communicate the number "1,000,000"—only 7 (or 20 if you're working base-2). For example, in an instance with n items, knapsack capacity 2^n, and all item values and sizes at most 2^n, the input size is $O(n^2)$—$O(n)$ numbers with $O(n)$ digits each—while the running time of the dynamic programming algorithm is exponentially larger (proportional to $n \cdot 2^n$).

The algorithmic strategy of designing fast and correct algorithms (for special cases) uses the entire algorithmic toolbox that we developed in *Parts 1–3*. For this reason, no chapter of this book is dedicated to this strategy. We will, however, encounter along the way further examples of polynomial-time solvable special cases of NP-hard problems, including the traveling salesman, satisfiability, and graph coloring problems (see Problems 19.8 and 21.12).

19.4.3   Compromising on Correctness

The second algorithmic strategy, which is particularly popular in time-critical applications, is to insist on generality and speed at the expense of correctness. Algorithms that are not always correct are sometimes called *heuristic algorithms*.[21]

Ideally, a heuristic algorithm is "mostly correct." This could mean one or both of two things:

Relaxations of Correctness

1. The algorithm is correct on "most" inputs.[22]

2. The algorithm is "almost correct" on every input.

---

[21]In *Parts 1–3*, there is exactly one example of a mostly-but-not-always-correct solution: bloom filters, a small-space data structure that supports super-fast insertions and lookups, at the expense of occasional false positives.

[22]For example, one typical implementation of a bloom filter has a 2% false positive rate, with 98% of lookups answered correctly.

The second property is easiest to interpret for optimization problems, in which the goal is to compute a feasible solution (like a traveling salesman tour) with the best objective function value (like the minimum total cost). "Almost correct" then means that the algorithm outputs a feasible solution with objective function value close to the best possible, like a traveling salesman tour with total cost not much more than that of an optimal tour.

Your existing algorithmic toolbox for designing fast exact algorithms is directly useful for designing fast heuristic algorithms. For example, Sections 20.1–20.3 describe greedy heuristics for problems ranging from scheduling to influence maximization in social networks. These heuristic algorithms come with proofs of "approximate correctness" guaranteeing that, for every input, the objective function value of the algorithm's output is within a modest constant factor of the best-possible objective function value.[23]

Sections 20.4–20.5 augment your toolbox with the *local search* algorithm design paradigm. Local search and its generalizations are unreasonably effective in practice at tackling many NP-hard problems, including the TSP, even though local search algorithms rarely possess compelling approximate correctness guarantees.

19.4.4   Compromising on Worst-Case Running Time

The final strategy is appropriate for applications in which you cannot afford to compromise on correctness and are therefore unwilling to consider heuristic algorithms. Every correct algorithm for an NP-hard problem must run in super-polynomial time on some inputs (assuming the P $\neq$ NP conjecture). The goal, therefore, is to design an algorithm that is as fast as possible—at a minimum, one that improves dramatically on naive exhaustive search. This could mean one or both of two things:

Relaxations of Polynomial Running Time

1. The algorithm typically runs quickly (for example, in polynomial time) on the inputs that are relevant to

---

[23]Some authors call such algorithms "approximation algorithms" while reserving the term "heuristic algorithms" for algorithms that lack such proofs of approximate correctness.

> your application.
>
> 2. The algorithm is faster than exhaustive search on every
>   input.

In the second case, we should still expect the algorithm to run in exponential time on some inputs—after all, the problem is NP-hard. For example, Section 21.1 employs dynamic programming to beat exhaustive search for the TSP, reducing the running time from $O(n!)$ to $O(n^2 \cdot 2^n)$, where n is the number of vertices. Section 21.2 combines randomization with dynamic programming to beat exhaustive search for the problem of finding long paths in graphs (with running time $O((2e)^k \cdot m)$ rather than $O(n^k)$, where n and m denote the number of vertices and edges in the input graph, k the target path length, and $e = 2.718\ldots$).

Making progress on relatively large instances of NP-hard problems typically requires additional tools that do not possess better-than-exhaustive-search running time guarantees but are unreasonably effective in many applications. Sections 21.3–21.5 outline how to stand on the shoulders of experts who, over several decades, have developed remarkably potent solvers for mixed integer programming ("MIP") and satisfiability ("SAT") problems. Many NP-hard optimization problems (such as the TSP) can be encoded as mixed integer programming problems. Many NP-hard feasibility-checking problems (such as checking for a conflict-free assignment of classes to classrooms) are easily expressed as satisfiability problems. Whenever you face an NP-hard problem that can be easily specified as a MIP or SAT problem, try applying the latest and greatest solvers to it. There's no guarantee that a MIP or SAT solver will solve your particular instance in a reasonable amount of time—the problem is NP-hard, after all—but they constitute cutting-edge technology for tackling NP-hard problems in practice.

19.4.5   Key Takeaways

If you're shooting for level-1 knowledge of NP-hardness (Section 19.2), the most important things to remember are:

> ## Three Facts About NP-Hard Problems
>
> 1. *Ubiquity:* Practically relevant NP-hard problems are everywhere.
>
> 2. *Intractability:* Under a widely believed mathematical conjecture, no NP-hard problem can be solved by any algorithm that is always correct and always runs in polynomial time.
>
> 3. *Not a death sentence:* NP-hard problems can often (but not always) be solved in practice, at least approximately, through sufficient investment of resources and algorithmic sophistication.

19.5   Proving NP-Hardness: A Simple Recipe

How can you recognize NP-hard problems when they arise in your own work, so that you can adjust your ambitions accordingly and abandon the search for an algorithm that is general-purpose, correct, and fast? Nobody wins if you spend weeks or months of your life inadvertently trying to refute the $P \neq NP$ conjecture.

First, know a collection of simple and common NP-hard problems (like the 19 problems in Chapter 22); in the simplest scenario, your application will literally boil down to one of these problems. Second, sharpen your ability to spot reductions between computational problems. Reducing one problem to another can spread computational tractability from the latter to the former. Turning this statement on its head, such a reduction can also spread computational *intractability* in the opposite direction, from the former problem to the latter. Thus, to show that a computational problem that you care about is NP-hard, all you need to do is reduce a known NP-hard problem to it.

The rest of this section elaborates on these points and provides one simple example; for a deep dive, see Chapter 22.

19.5.1   Reductions

Any problem B that is at least as hard as an NP-hard problem A is itself NP-hard. The phrase "at least as hard as" can be formalized using *reductions*.

Reductions

A problem *A* *reduces* to another problem *B* if an algorithm that solves *B* can be easily translated into one that solves *A* (Figure 19.1).

When discussing NP-hard problems, "easily translate" means that problem *A* can be solved using at most a polynomial (in the input size) number of invocations of a subroutine that solves problem *B*, along with a polynomial amount of additional work (outside of the subroutine calls).

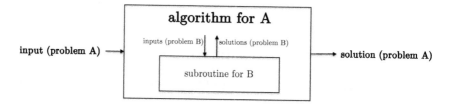

Figure 19.1: If the problem *A* reduces to the problem *B*, then *A* can be solved using a polynomial (in the input size) number of calls to a subroutine for *B*, plus a polynomial amount of additional work.

19.5.2   Using Reductions to Design Fast Algorithms

Seasoned algorithm designers are always on the lookout for reductions—why solve a problem from scratch if you don't have to? Examples from *Parts 1–3* related to the problems listed in Section 19.3.1 include:

Familiar Examples of Reductions

1. Finding the median of an array of integers reduces to the problem of sorting the array. (After sorting the array, return the middle element.)

2. The all-pairs shortest path problem reduces to the single-source shortest path problem. (Invoke a single-

> source shortest-path algorithm once with each possible choice of a starting vertex in the input graph.)
>
> 3. The longest common subsequence problem reduces to the sequence alignment problem. (Invoke a sequence alignment algorithm with the two input strings, a penalty of 1 per inserted gap, and a very large penalty for each mismatch of two different symbols.)[24]

These reductions take after the light side of the force and serve the honorable mission of creating new fast algorithms from old ones, thereby advancing the frontier of computational tractability. For example, the first reduction translates the `MergeSort` algorithm into an $O(n \log n)$-time median-finding algorithm or, more generally, any $T(n)$-time sorting algorithm into an $O(T(n))$-time median-finding algorithm, where n is the array length. The second reduction translates any $T(m, n)$-time algorithm for the single-source shortest path problem into an $O(n \cdot T(m, n))$-time algorithm for the all-pairs shortest path problem, where m and n denote the number of edges and vertices, respectively; and the third a $T(m, n)$-time algorithm for the sequence alignment problem into an $O(T(m, n))$-time algorithm for the longest common subsequence problem, where m and n denote the lengths of the two input strings.

Quiz 19.3

Suppose that a problem A can be solved by invoking a subroutine for a problem B at most $T_1(n)$ times and performing at most $T_2(n)$ additional work (outside of the subroutine calls), where n denotes the input size. Provided with a subroutine that solves problem B in time at most $T_3(n)$ on size-n inputs, how much time do you need to solve problem A? (Choose the strongest true statement. Assume that a program must use at least s primitive operations to construct a size-s input to a subroutine call.)

---

[24]Recall that an instance of the sequence alignment problem is specified by two strings over some alphabet Σ (like $\{A, C, G, T\}$), a penalty α_{xy} for each symbol pair $x, y \in \Sigma$, and a nonnegative gap penalty α_{gap}. The goal is to compute an alignment of the input strings with the minimum-possible total penalty.

a) $T_1(n) + T_2(n) + T_3(n)$

b) $T_1(n) \cdot T_2(n) + T_3(n)$

c) $T_1(n) \cdot T_3(n) + T_2(n)$

d) $T_1(n) \cdot T_3(T_2(n)) + T_2(n)$

(See Section 19.5.5 for the solution and discussion.)

Quiz 19.3 shows that whenever a problem A reduces to another problem B, any polynomial-time algorithm for B can be translated into one for A:[25]

Reductions Spread Tractability

If problem A reduces to problem B and B can be solved by a polynomial-time algorithm, then A can also be solved by a polynomial-time algorithm (Figure 19.2).

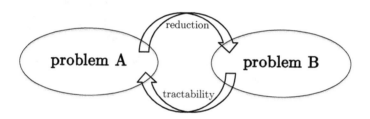

Figure 19.2: Spreading tractability from B to A: If problem A reduces to problem B and B is computationally tractable, then A is also computationally tractable.

19.5.3   Using Reductions to Spread NP-Hardness

The theory of NP-hardness follows the dark side of the force, ne-fariously using reductions to spread the curse of computational in-tractability (in the opposite direction of Figure 19.2). Let's turn the

---

[25]If the functions $T_1(n)$, $T_2(n)$, and $T_3(n)$ in Quiz 19.3 are each bounded by a polynomial function of n, so are their sums, products, and compositions. For example, if $T_1(n) \le a_1 n^{d_1}$ and $T_2(n) \le a_2 n^{d_2}$, where a_1, a_2, d_1, and d_2 are positive constants (independent of n), then $T_1(n) \cdot T_2(n) \le (a_1 a_2) n^{(d_1 + d_2)}$ and $T_1(T_2(n)) \le (a_1 a_2^{d_1}) n^{(d_1 d_2)}$.

preceding boxed statement on its head. Suppose that a problem A reduces to another problem B. Suppose further that A is NP-hard, meaning that a polynomial-time algorithm for A would refute the P $\neq$ NP conjecture. Well, a polynomial-time algorithm for B would automatically lead to one for A (because A reduces to B); this, in turn, would refute the P $\neq$ NP conjecture. In other words, B is also NP-hard!

Reductions Spread Intractability

If problem A reduces to problem B and A is NP-hard, then B is also NP-hard (Figure 19.3).

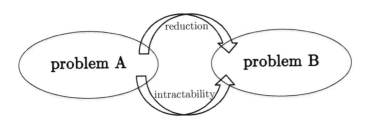

Figure 19.3: Spreading intractability in the opposite direction, from A to B: If problem A reduces to problem B and A is computationally intractable, then B is also computationally intractable.

We therefore have a remarkably simple two-step recipe for proving that a problem is NP-hard:

How to Prove a Problem Is NP-Hard

To prove that a problem B is NP-hard:

1. Choose an NP-hard problem A.

2. Prove that A reduces to B.

Carrying out the first step requires knowledge of some known NP-hard problems; Chapter 22 will get you started. The second step builds on your already-developed skills in finding reductions between problems; these will be honed further through practice in Chapter 22. Let's get the gist of how this recipe works by revisiting a familiar problem:

the single-source shortest path problem, with negative edge lengths allowed.

19.5.4 NP-Hardness of Cycle-Free Shortest Paths

In the *single-source shortest path problem*, the input consists of a directed graph $G = (V, E)$, a real-valued length ℓ_e for each edge $e \in E$, and a starting vertex $s \in V$. The *length* of a path is the sum of the lengths of its edges. The goal is to compute, for every possible destination $v \in V$, the minimum length $dist(s, v)$ of a directed path in G from s to v. (If no such path exists, $dist(s, v)$ is defined as $+\infty$.) Importantly, negative edge lengths are allowed.[26,27] For example, the shortest-path distances from s in the graph

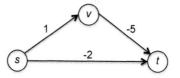

are $dist(s, s) = 0$, $dist(s, v) = 1$, and $dist(s, t) = -4$.

Negative Cycles

How should we define shortest-path distances in a graph like the following?

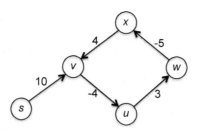

---

[26] Remember that paths in a graph might represent abstract sequences of decisions rather than something physically realizable. For example, if you want to compute a profitable sequence of financial transactions involving both buying and selling, you're looking for a shortest path in a graph with edge lengths that are both positive and negative.

[27] In graphs with only nonnegative edge lengths, the single-source shortest path problem can be solved in blazingly fast fashion by Dijkstra's algorithm (see Chapter 9 of *Part 2*).

This graph has a *negative cycle*, meaning a directed cycle for which the sum of the edge lengths is negative. There is a one-hop *s-v* path with length 10. Tacking a cycle traversal at the end produces a five-hop *s-v* path with total length 8. Adding a second traversal decreases the overall length to 6, and so on. If we allow paths with cycles, then this graph has no shortest *s-v* path.

The Cycle-Free Shortest Path Problem

An obvious alternative is to forbid paths with cycles, insisting that every vertex is visited at most once.

Problem: Cycle-Free Shortest Paths (CFSP)

Input: A directed graph $G = (V, E)$, a starting vertex $s \in V$, and a real-valued length ℓ_e for each edge $e \in E$.

Output: For every $v \in V$, the minimum length of a cycle-free *s-v* path in G (or $+\infty$, if there is no *s-v* path in G).

Unfortunately, this version of the problem is NP-hard.[28]

Theorem 19.1 (NP-Hardness of Cycle-Free Shortest Paths)
The cycle-free shortest path problem is NP-hard.

On Lemmas, Theorems, and the Like

In mathematical writing, the most important technical statements are labeled *theorems*. A *lemma* is a technical statement that assists with the proof of a theorem (much as a subroutine assists with the implementation of a larger program). A *corollary* is a statement that follows immediately from an already-

---

[28]This explains why the Bellman-Ford algorithm (see Chapter 18 of *Part 3*)—along with every other polynomial-time shortest-path algorithm—solves only a special case of the problem (input graphs without negative cycles, in which shortest paths are automatically cycle-free). Theorem 19.1 shows that, assuming the P ≠ NP conjecture, no such algorithm can compute correct cycle-free shortest-path distances in general.

proven result, such as a special case of a theorem. We use the term *proposition* for stand-alone technical statements that are not particularly important in their own right.

The Directed Hamiltonian Path Problem

We can prove Theorem 19.1 by following the two-step recipe in Section 19.5.3. For the first step, we'll use a famous NP-hard problem known as the *directed Hamiltonian path* problem.

Problem: Directed Hamiltonian Path (DHP)

Input: A directed graph $G = (V, E)$, a starting vertex $s \in V$, and an ending vertex $t \in V$.

Output: "Yes" if G contains an s-t path visiting every vertex $v \in V$ exactly once (called an s-t *Hamiltonian path*), and "no" otherwise.

For example, of the two directed graphs

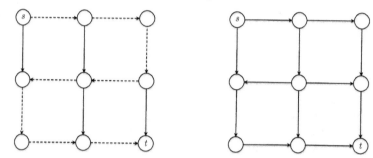

the first has an s-t Hamiltonian path (the dashed edges) while the second does not.

Proof of Theorem 19.1

Section 22.6 proves that the directed Hamiltonian path problem is NP-hard (again using the two-step recipe in Section 19.5.3). For now, we'll take its NP-hardness on faith and move on to the second step of the recipe, in which we reduce a known NP-hard problem (in this case, directed Hamiltonian path) to the problem of interest (cycle-free shortest paths).

Lemma 19.2 (Reduction from DHP to CFSP) *The directed Hamiltonian path problem reduces to the cycle-free shortest path problem.*

Proof: How can we use a subroutine for the cycle-free shortest path problem to solve the directed Hamiltonian path problem (recall Figure 19.1)? Suppose we're given an instance of the latter problem, specified by a directed graph $G = (V, E)$, a starting vertex $s \in V$, and an ending vertex $t \in V$. The assumed cycle-free shortest path subroutine is awaiting a graph (and we have one to offer, our own input graph G) and a starting vertex s (ditto). It's not prepared for an ending vertex, but we can keep mum about t. The subroutine is expecting to receive real-valued edge lengths, however, so we'll have to make some up. We can trick the subroutine into thinking that long paths (like an s-t Hamiltonian path) are actually short by giving each edge a negative length. Summarizing, the reduction is (Figure 19.4):

---

1. Assign every edge $e \in E$ a length $\ell_e = -1$.

2. Compute cycle-free shortest paths using the assumed subroutine, reusing the same input graph G and starting vertex s.

3. If the length of a shortest cycle-free path from s to t is $-(|V| - 1)$, return "yes." Otherwise, return "no."

To prove that this reduction is correct, we must show that it returns "yes" whenever the input graph G contains an s-t Hamiltonian path, and "no" otherwise. In the constructed cycle-free shortest paths instance, the minimum length of a cycle-free s-t path equals -1 times the maximum number of hops in a cycle-free s-t path of the original input graph G. A cycle-free s-t path uses $|V| - 1$ hops if it's an s-t Hamiltonian path (to visit all $|V|$ vertices), and fewer otherwise. So, if G has an s-t Hamiltonian path, the cycle-free shortest-path distance from s to t in the constructed instance is $-(|V| - 1)$; otherwise, it is

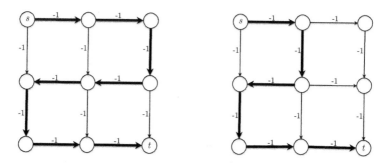

Figure 19.4: Example of the reduction in the proof of Lemma 19.2. The
s-t Hamiltonian path in the first graph translates to a cycle-free s-t path
with length -8. The second graph has no s-t Hamiltonian path, and the
minimum length of a cycle-free s-t path is -6.

bigger (that is, less negative). Either way, the reduction returns the
correct answer. $\mathcal{QED}$[29]

By the two-step recipe, Lemma 19.2 and the NP-hardness of the
directed Hamiltonian path problem prove Theorem 19.1. Chapter 22
presents many more examples of this recipe in action.

19.5.5   Solution to Quiz 19.3

Correct answer: (d). At first blush, the answer seems to be (c):
Each of the at most $T_1(n)$ calls to the subroutine performs at
most $T_3(n)$ operations; beyond these calls, the algorithm performs
at most $T_2(n)$ operations, for an overall running time of at most
$T_1(n) \cdot T_3(n) + T_2(n)$.

This reasoning is correct for most natural reductions between
problems, including the three examples in Section 19.5.2. Technically,
however, a reduction might, given a size-n input, invoke a subroutine
for B on inputs *larger* than n. For example, imagine a reduction that
takes as input a graph and, for whatever reason, adds some additional
vertices or edges to it before invoking the subroutine for B. What's
the worst that could happen? Because the reduction performs at
most $T_2(n)$ operations outside of the subroutine calls, it has time only

---

[29]"Q.e.d." is an abbreviation for *quod erat demonstrandum* and means "that
which was to be demonstrated." In mathematical writing, it is used at the end of
a proof to mark its completion.

to write down inputs of problem B of size at most $T_2(n)$. Thus, each of the $T_1(n)$ invocations of B requires at most $T_3(T_2(n))$ operations, for an overall running time of $T_1(n) \cdot T_3(T_2(n)) + T_2(n)$.

19.6   Rookie Mistakes and Acceptable Inaccuracies

NP-hardness is a pretty technical topic but, at the same time, highly relevant for practicing algorithm designers and serious programmers. Outside of textbooks and research papers, computer scientists often take liberties with the precise mathematical definitions in the interest of easier communication. Some types of inaccuracies will mark you as a clueless newbie, while others are culturally acceptable. How would you ever know which is which? Because I'm going to tell you, right now.

Rookie Mistake #1

Thinking that "NP" stands for "not polynomial."

You don't need to remember what "NP" actually stands for as long as you avoid this rookie mistake.[30]

Rookie Mistake #2

Saying that a problem is "an NP problem" or "in NP" instead of "NP-hard."

Readers who persevere through Section 23.3 will learn that being "an NP problem" or "in NP" is actually a *good* thing, not a bad thing.[31] So don't forget the "-hard" after the "NP."

Rookie Mistake #3

Thinking that NP-hardness doesn't matter because NP-hard problems can generally be solved in practice.

---

[30] So, what *does* it stand for? Section 23.3 provides the historical context, but in case the suspense is killing you... "nondeterministic polynomial time."

[31] Specifically, it means that if someone handed you a solution on a silver platter (like a completed Sudoku puzzle), you could verify its validity in polynomial time.

It's true that NP-hardness is not a death sentence and that NP-hard problems have been tamed, using sufficient human and computational investment, in many practical applications; see Chapter 24 for an in-depth case study. But there are plenty of other applications in which computational problems have been modified or even abandoned because of the challenges posed by NP-hardness. (Naturally, people report their successes in tackling NP-hard problems much more eagerly than their failures!) If it really were true that no problems are hard in practice, why would heuristic algorithms be so common? For that matter, how could modern ecommerce even exist?[32]

Rookie Mistake #4

Thinking that advances in computer technology will rescue us from NP-hardness.

Moore's Law and correspondingly larger input sizes only exacerbate the issue, with an increasingly big gulf between running times that are polynomial and those that are not (Section 19.3.2). Quantum computers enable algorithms that improve on exhaustive search but appear inadequate for solving any NP-hard problem in polynomial time (Section 19.3.8).

Rookie Mistake #5

Devising a reduction in the wrong direction.

A reduction from one problem A to another problem B spreads NP-hardness from A to B, not the other way around (compare Figures 19.2 and 19.3). Because we're so accustomed to designing reductions that spread tractability rather than intractability, this is the hardest mistake to avoid. Whenever you think you've proved that a problem is NP-hard, go back and triple-check that your reduction goes in the correct direction—the same direction in which you're attempting to spread intractability.

---

[32]Ecommerce relies on cryptosystems like RSA, the security of which depends on the computational intractability of factoring large integers. A polynomial-time algorithm for any NP-hard problem would, via reductions, immediately lead to a polynomial-time factoring algorithm.

Acceptable Inaccuracies

Next are three statements that are culturally acceptable despite being unproven or technically incorrect. None of these will shake anyone's confidence in your understanding of NP-hardness.

Acceptable Inaccuracy #1

Assuming that the P $\neq$ NP conjecture is true.

The status of the P $\neq$ NP conjecture remains open, though most experts are believers. While we wait for our mathematical understanding to catch up to our intuition, many treat the conjecture as a law of nature.

Acceptable Inaccuracy #2

Using the terms "NP-hard" and "NP-complete" interchangeably.

"NP-completeness" is a specific type of NP-hardness; the details are technical and deferred to Section 23.3. The algorithmic implications are the same either way: Whether NP-complete or NP-hard, the problem is not polynomial-time solvable (assuming the P $\neq$ NP conjecture).

Acceptable Inaccuracy #3

Conflating NP-hardness with requiring exponential time in the worst case.

This is the oversimplified interpretation of NP-hardness from the beginning of Section 19.3. This conflation is technically inaccurate (see Section 19.3.9) but faithful to how most experts think about NP-hardness; no one will bat an eye if you make it yourself.

The Upshot

☆ A polynomial-time algorithm is one with worst-case running time $O(n^d)$, where n denotes the input size and d is a constant.

☆ A computational problem is polynomial-time solvable if there is a polynomial-time algorithm that solves it correctly for every input.

☆ The theory of NP-hardness equates "easy" with polynomial-time solvable. Oversimplifying, a "hard" problem is one requiring exponential time to solve in the worst case.

☆ Informally, the $P \neq NP$ conjecture asserts that checking a solution to a problem can be easier than coming up with your own from scratch.

☆ Provisionally, a computational problem is NP-hard if a polynomial-time algorithm solving it would refute the $P \neq NP$ conjecture.

☆ A polynomial-time algorithm for any NP-hard problem would automatically solve thousands of problems that have resisted the efforts of countless brilliant minds over many decades.

☆ NP-hard problems are ubiquitous.

☆ To make progress on an NP-hard problem, the algorithm designer must compromise on generality, correctness, or speed.

☆ Fast heuristic algorithms run quickly but are not always correct. The greedy and local search paradigms are particularly useful for designing such algorithms.

☆ Dynamic programming can improve on exhaustive search for several NP-hard problems.

☆ Mixed integer programming and satisfiability solvers constitute cutting-edge technology for tackling NP-hard problems in practice.

☆ Problem A reduces to problem B if A can be solved using a polynomial number of calls to a subroutine solving B and a polynomial amount of additional work.

☆ Reductions spread tractability: If problem A reduces to problem B and B can be solved by a polynomial-time algorithm, then A can also be solved by a polynomial-time algorithm.

☆ Reductions spread intractability, in the opposite direction: If problem A reduces to problem B and A is NP-hard, then B is also NP-hard.

☆ To prove that a problem B is NP-hard: (i) choose an NP-hard problem A; (ii) prove that A reduces to B.

Test Your Understanding

Problem 19.1 *(S)* Suppose that a computational problem B that you care about is NP-hard. Which of the following are true? (Choose all that apply.)

a) NP-hardness is a "death sentence"; you shouldn't bother trying to solve the instances of B that are relevant for your application.

b) If your boss criticizes you for failing to find a polynomial-time algorithm for B, you can legitimately respond that thousands of brilliant minds have likewise tried and failed to solve B.

c) You should not try to design an algorithm that is guaranteed to solve B correctly and in polynomial time for every possible instance of the problem (unless you're explicitly trying to refute the $P \neq NP$ conjecture).

d) Because the dynamic programming paradigm is useful only for designing exact algorithms, there's no point in trying to apply it to problem B.

Problem 19.2 *(S)* Which of the following statements are true? (Choose all that apply.)

a) The MST problem is computationally tractable because the number of spanning trees of a graph is polynomial in the number n of vertices and the number m of edges.

b) The MST problem is computationally tractable because there are at most m possibilities for the total cost of a spanning tree of a graph.

c) Exhaustive search does not solve the TSP in polynomial time because a graph has an exponential number of traveling salesman tours.

d) The TSP is computationally intractable because a graph has an exponential number of traveling salesman tours.

Problem 19.3 *(S)* Which of the following statements are true? (Choose all that apply.)

a) If the P $\neq$ NP conjecture is true, NP-hard problems can never be solved in practice.

b) If the P $\neq$ NP conjecture is true, no NP-hard problem can be solved by an algorithm that is always correct and that always runs in polynomial time.

c) If the P $\neq$ NP conjecture is false, NP-hard problems can always be solved in practice.

d) If the P $\neq$ NP conjecture is false, some NP-hard problems are polynomial-time solvable.

Problem 19.4 *(S)* Which of the following statements are implied by the P $\neq$ NP conjecture? (Choose all that apply.)

a) Every algorithm that solves an NP-hard problem runs in superpolynomial time in the worst case.

b) Every algorithm that solves an NP-hard problem runs in exponential time in the worst case.

c) Every algorithm that solves an NP-hard problem always runs in super-polynomial time.

d) Every algorithm that solves an NP-hard problem always runs in exponential time.

Problem 19.5 *(S)* Suppose that a problem A reduces to another problem B. Which of the following statements are always true? (Choose all that apply.)

a) If A is polynomial-time solvable, then B is also polynomial-time solvable.

b) If B is NP-hard, then A is also NP-hard.

c) B also reduces to A.

d) B cannot reduce to A.

e) If the problem B reduces to another problem C, then A also reduces to C.

Problem 19.6 *(S)* Assume that the P $\neq$ NP conjecture is true. Which of the following statements about the knapsack problem (Section 19.4.2) are correct? (Choose all that apply.)

a) The special case in which all item sizes are positive integers less than or equal to n^5, where n is the number of items, can be solved in polynomial time.

b) The special case in which all item values are positive integers less than or equal to n^5, where n is the number of items, can be solved in polynomial time.

c) The special case in which all item values, all item sizes, and the knapsack capacity are positive integers can be solved in polynomial time.

d) There is no polynomial-time algorithm for the knapsack problem in general.

Challenge Problems

Problem 19.7 *(H)* The input in the *traveling salesman path problem*
(TSPP) is the same as that in the TSP, and the goal is to compute a
minimum-cost cycle-free path that visits every vertex (that is, a tour
without its final edge). Prove that the TSPP reduces to the TSP and
vice versa.

Problem 19.8 *(H)* This problem describes a computationally
tractable special case of the TSP. Consider a connected and acyclic
graph $T = (V, F)$ in which each edge $e \in F$ has a nonnegative length
$a_e \geq 0$. Define the corresponding *tree instance* $G = (V, E)$ of the TSP
by setting the cost c_{vw} of each edge $(v, w) \in E$ equal to the length
$\sum_{e \in P_{vw}} a_e$ of the (unique) v-w path P_{vw} in T. For example:

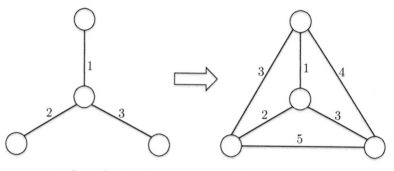

connected acyclic graph   corresponding tree instance of TSP

Design a linear-time algorithm that, given a connected acyclic graph
with nonnegative edge lengths, outputs a minimum-cost traveling
salesman tour of the corresponding tree instance. Prove that your
algorithm is correct.

Programming Problems

Problem 19.9 Implement in your favorite programming language
the exhaustive search algorithm for the TSP (as seen in Quiz 19.2).
Give your implementation a spin on instances with edge costs chosen
independently and uniformly at random from the set $\{1, 2, \ldots, 100\}$.
How large an input size (that is, how many vertices) can your program
reliably process in under a minute? What about in under an hour?
(See www.algorithmsilluminated.org for test cases and challenge
data sets.)

Chapter 20

Compromising on Correctness: Efficient Inexact Algorithms

You can't have it all with NP-hard problems and must give up on generality, correctness, or speed. When generality and speed are mission-critical, it's time to consider heuristic algorithms that are not always correct. The goal is then to minimize the damage and design a general-purpose and fast algorithm that is—perhaps provably, or at least empirically—"approximately correct." This chapter illustrates through examples how to use techniques both new (like local search) and old (like greedy algorithms) for this purpose. The case studies concern scheduling (Section 20.1), team selection (Section 20.2), social network analysis (Section 20.3), and the TSP (Section 20.4).

20.1 Makespan Minimization

Our first case study concerns *scheduling* and the goal of assigning tasks to shared resources to optimize some objective. For example, a resource could represent a computer processor (with tasks corresponding to jobs), a classroom (with tasks corresponding to lectures), or a workday (with tasks corresponding to meetings).

20.1.1 Problem Definition

In scheduling problems, the tasks to be completed are usually called *jobs* and the resources are called *machines*. A *schedule* specifies, for each job, one machine to process it. There are a lot of possible schedules. Which one should we prefer?

Suppose that each job j has a known *length* ℓ_j, which is the amount of time required to process it (for example, the length of a lecture or meeting). We'll consider one of the most common objectives in applications, of scheduling the jobs so that they all complete as quickly

41

as possible. The following *objective function* formalizes this idea by assigning a numerical score to every schedule and quantifying what we want:

The Makespan of a Schedule

1. The *load* of a machine in a schedule is the sum of the lengths of the jobs assigned to it.

2. The *makespan* of a schedule is the maximum of the machine loads.

Machine loads and the makespan are the same no matter how jobs are ordered on each machine, so schedules specify only assignments of jobs to machines and not orderings of jobs.

Quiz 20.1

What are the makespans of the following schedules? (Jobs are labeled with their lengths.)

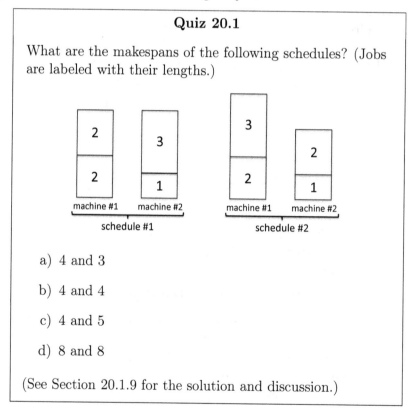

a) 4 and 3

b) 4 and 4

c) 4 and 5

d) 8 and 8

(See Section 20.1.9 for the solution and discussion.)

An "optimal" schedule is then one with the minimum-possible makespan. For example, in Quiz 20.1, the first schedule is the unique one that minimizes the makespan.

Problem: Makespan Minimization

Input: A set of n jobs with positive lengths $\ell_1, \ell_2, \ldots, \ell_n$ and m identical machines.

Output: An assignment of jobs to machines that minimizes the makespan.

For example, if jobs represent parts of a computational task to be processed in parallel (such as the jobs that make up a MapReduce or Hadoop program), the schedule's makespan governs when the entire computation completes.

Minimizing the makespan is an NP-hard problem (see Problem 22.10). Could there be an algorithm that is general-purpose, fast, and "almost correct"?

20.1.2   Greedy Algorithms

For many computational problems (both easy and hard), greedy algorithms are a great place to start brainstorming. To review (for example, from Chapter 13 of *Part 3*), the greedy algorithm design paradigm is:

The Greedy Paradigm

Construct a solution iteratively, via a sequence of myopic decisions, and hope that everything works out in the end.

The two biggest selling points of greedy algorithms are that they're usually easy to come up with and they tend to be very fast. The downside is that most greedy algorithms return an incorrect solution in some cases. But for an NP-hard problem, this flaw is shared by all fast algorithms: *No* polynomial-time algorithm is correct on all inputs (assuming, as usual, that the P $\neq$ NP conjecture is true)! Thus, the greedy paradigm is particularly apropos for the design of fast heuristic algorithms for NP-hard problems, and it plays a starring role in this chapter.

20.1.3   Graham's Algorithm

What would a greedy algorithm look like for the makespan minimization problem? Perhaps the simplest approach would be a single-pass algorithm, which assigns jobs irrevocably to machines one by one. To which machine should a job be assigned? Because we're after the most balanced schedule possible, the obvious greedy strategy is to assign a job to the machine that can best tolerate it—the machine with the *smallest* current load. This greedy algorithm is known as *Graham's algorithm*.[1]

Graham

Input: a set $\{1, 2, \ldots, m\}$ of machines and a set $\{1, 2, \ldots, n\}$ of jobs with positive lengths $\ell_1, \ell_2, \ldots, \ell_n$.
Output: an assignment of jobs to machines.

```
// Initialization
1 for i = 1 to m do
2     J_i := ∅              // jobs assigned to machine i
3     L_i := 0              // current load of machine i
// Main loop
4 for j = 1 to n do
5     k := argmin_{i=1}^{m} L_i      // least-loaded machine²
6     J_k := J_k ∪ {j}               // assign current job
7     L_k := L_k + ℓ_j               // update loads
8 return J_1, J_2, ..., J_m
```

On Pseudocode

This book series explains algorithms using a mixture of high-level pseudocode and English (as above). I'm

---

[1] Proposed by Ronald L. Graham in the paper "Bounds on Multiprocessing Time Anomalies" (*SIAM Journal on Applied Mathematics*, 1969).

[2] For a sequence $a_1, a_2, \ldots, a_n$ of real numbers, $\operatorname{argmin}_{i=1}^{n} a_i$ denotes the index of the smallest number. (If multiple numbers are tied for the smallest, interpret $\operatorname{argmin}_{i=1}^{n} a_i$ as breaking ties between them arbitrarily.) The function $\operatorname{argmax}_{i=1}^{n} a_i$ is defined similarly.

assuming that you have the skills to translate such high-level descriptions into working code in your favorite programming language. Several other books and resources on the Web offer concrete implementations of various algorithms in specific programming languages.

The first benefit of emphasizing high-level descriptions over language-specific implementations is flexibility. While I assume familiarity with *some* programming language, I don't care which one. Second, this approach promotes the understanding of algorithms at a deep and conceptual level, unencumbered by low-level details. Seasoned programmers and computer scientists generally think and communicate about algorithms at a similarly high level.

Still, there is no substitute for the detailed understanding of an algorithm that comes from providing your own working implementation of it. I strongly encourage you to implement as many of the algorithms in this book as you have time for. (It's also a great excuse to pick up a new programming language!) For guidance, see the end-of-chapter Programming Problems and supporting test cases.

20.1.4   Running Time

Is Graham's algorithm any good? As usual with greedy algorithms, its running time is easy to analyze. If the argmin computation in line 5 is implemented by exhaustive search through the m possibilities, each of the n iterations of the main loop runs in $O(m)$ time (implementing the J_i's as linked lists, for example). Because only $O(m)$ work is performed outside the main loop, this straightforward implementation leads to a running time of $O(mn)$.

Readers who have experience with data structures should recognize an opportunity for improvement. The work done by the algorithm boils down to repeated minimum computations, so a light bulb should go off in your head: This algorithm calls out for a heap data structure![3] Because a heap reduces the running time of a minimum computation from linear to logarithmic, its use here leads to a blazingly fast $O(n \log m)$-time implementation of the Graham algorithm. Problem 20.6 asks you to fill in the details.

20.1.5 Approximate Correctness

What about the makespan of the schedule constructed by Graham's algorithm?

Quiz 20.2

Suppose there are five machines and the list of jobs consists of twenty jobs with length 1 each, followed by a single job with length 5. What is the makespan of the schedule output by the Graham algorithm, and what is the smallest-possible makespan of a schedule of these jobs?

a) 5 and 4

b) 6 and 5

c) 9 and 5

d) 10 and 5

(See Section 20.1.9 for the solution and discussion.)

Quiz 20.2 demonstrates that the Graham algorithm does not always output an optimal schedule. This is no surprise, given that the problem is NP-hard and the algorithm runs in polynomial time. (If the algorithm *were* always correct, we would have refuted the P $\neq$ NP conjecture!) Even so, the example in Quiz 20.2 should give you pause. Could there be other, more complicated inputs for which Graham's algorithm performs still more poorly? Happily, examples of the type in Quiz 20.2 are as bad as it gets:

Theorem 20.1 (Graham: Approximate Correctness) *The makespan of the schedule output by the Graham algorithm is always at*

---

[3]See, for example, Chapter 10 of *Part 2*.

most $2 - \frac{1}{m}$ *times the minimum-possible makespan, where m denotes the number of machines.*[4,5]

Graham's algorithm is, therefore, an "approximately correct" algorithm for the makespan minimization problem. Think of Theorem 20.1 as an insurance policy. Even in the doomsday scenario of a contrived input like that in Quiz 20.2, the makespan of the algorithm's schedule is no more than double what you'd get by exhaustive search. For more realistic inputs, you should expect the Graham algorithm to overdeliver and achieve a makespan much closer to the minimum possible; see also Problem 20.1.

The next section provides the full proof of Theorem 20.1. The time-constrained or math-phobic reader might prefer some brief but accurate intuition:

Intuition for Theorem 20.1

1. The *smallest* machine load is at most the (equal) machine loads in a perfectly balanced schedule, which in turn is at most the minimum-possible makespan (as the best-case scenario is a perfectly balanced schedule).

2. By the Graham algorithm's greedy criterion, the largest and smallest machine loads differ by at most the length of a single job, which in turn is at most the minimum-possible makespan (as every job has to go somewhere).

3. Thus, the largest machine load in the algorithm's output is at most twice the minimum-possible makespan.

20.1.6   Proof of Theorem 20.1

For the formal proof, fix an instance comprising jobs with lengths $\ell_1, \ell_2, \ldots, \ell_n$ and m machines. Directly comparing the minimum-possible makespan M^* and the makespan M of the schedule output by the Graham algorithm would be messy. Instead, the analysis hinges

---

[4]To generalize the bad example in Quiz 20.2 to an arbitrary number m of machines, use $m(m-1)$ length-1 jobs followed by a single job with length m.

[5]The multiplier $2 - \frac{1}{m}$ is sometimes called the *approximation ratio* of the algorithm, which in turn is called a $(2 - \frac{1}{m})$-*approximation algorithm*.

on two easy-to-compute lower bounds on M^*—the maximum job length and the average machine load—that are easily related to M and ultimately show that $M \leq (2 - \frac{1}{m})M^*$.

The first lower bound on M^* is simple: Every job must go somewhere, so it's impossible to achieve a makespan smaller than a job length.

Lemma 20.2 (Lower Bound #1 on the Optimal Makespan)
If M^* denotes the minimum makespan of any schedule and j a job,

$$M^* \geq \ell_j. \tag{20.1}$$

More generally, in every schedule, every job j is assigned to exactly one machine i and contributes ℓ_j to its load L_i. Thus, in every schedule, the sum of the machines' loads equals the sum of the jobs' lengths: $\sum_{i=1}^{m} L_i = \sum_{j=1}^{n} \ell_j$. In a perfect schedule, each machine has an *ideal* load, meaning an exact $\frac{1}{m}$ fraction of the total (that is, $\frac{1}{m} \sum_{j=1}^{n} \ell_j$). In any other schedule, some machines have a more-than-ideal load and others a less-than-ideal load. For example, in Quiz 20.1, in the first schedule both machines have ideal loads, while in the second schedule neither has an ideal load (with one over and the other under).

The second lower bound on M^* now follows from the fact that every schedule has a machine with a load equal to or larger than the ideal load:

Lemma 20.3 (Lower Bound #2 on the Optimal Makespan)
If M^* denotes the minimum makespan of any schedule, then

$$M^* \geq \underbrace{\frac{1}{m} \sum_{j=1}^{n} \ell_j}_{ideal\ load}. \tag{20.2}$$

The last step is to bound from above the makespan M of the Graham algorithm's schedule in terms of the two lower bounds introduced in Lemmas 20.2 and 20.3. Let i denote a machine with the largest load in this schedule (that is, with load L_i equal to M), and j the final job assigned to it (Figure 20.1(a)). Rewind the algorithm to the moment in time just before j's assignment and let $\widehat{L}_i$ denote i's load at that time. The new and final load L_i of the machine (and, hence, the makespan M) is $\ell_j + \widehat{L}_i$.

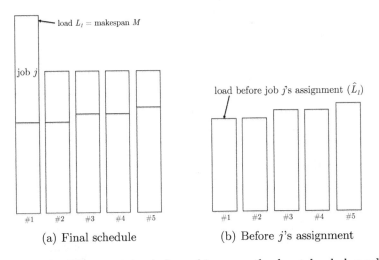

(a) Final schedule   (b) Before j's assignment

Figure 20.1: The most loaded machine was the least loaded machine immediately prior to its final job assignment.

How big could $\widehat{L}_i$ have been? By the greedy criterion of the Graham algorithm, i was the most lightly loaded machine at the time (Figure 20.1(b)). If the jobs $\{1, 2, \ldots, j-1\}$ prior to j were perfectly balanced across the machines, all machines' loads at that time would have been $\frac{1}{m} \sum_{h=1}^{j-1} \ell_h$; otherwise, the lightest load $\widehat{L}_i$ would have been even less. In any case, the final makespan $M = \ell_j + \widehat{L}_i$ is at most

$$\ell_j + \frac{1}{m} \sum_{h=1}^{j-1} \ell_h \leq \ell_j + \frac{1}{m} \sum_{h \neq j} \ell_h,$$

where on the right-hand side we have thrown in the missing (positive) terms $\ell_{j+1}/m, \ell_{j+2}/m, \ldots, \ell_n/m$ for convenience. Transferring ℓ_j/m from the first term to the second, we can write

$$M \leq \underbrace{\left(1 - \frac{1}{m}\right) \cdot \ell_j}_{\leq \left(1 - \frac{1}{m}\right) M^* \text{ by (20.1)}} + \underbrace{\frac{1}{m} \sum_{h=1}^{n} \ell_h}_{\leq M^* \text{ by (20.2)}} \leq \left(2 - \frac{1}{m}\right) \cdot M^*, \quad (20.3)$$

with the second inequality following from Lemma 20.2 (to bound the first term) and Lemma 20.3 (to bound the second term). This completes the proof of Theorem 20.1. $\mathcal{QED}$

20.1.7 Longest Processing Time First (LPT)

An insurance policy like the approximate correctness guarantee in Theorem 20.1 is reassuring, but it remains our duty to ask: Can we do better? Can we devise a different fast heuristic algorithm that is "even less incorrect," offering an insurance policy with a lower deductible? We can, using a familiar for-free primitive.

For-Free Primitives

You can think of an algorithm with a linear or near-linear running time as a primitive that can be used essentially "for free"—the amount of time required barely exceeds what you need to read the input. When you have such a blazingly fast primitive that is relevant to your problem, why not use it? For example, you can always sort your data in a preprocessing step, even if you're not quite sure how it will help later. One of the goals of this book series is to stock your algorithmic toolbox with as many for-free primitives as possible, ready to be applied at will.

What goes wrong with the Graham algorithm in the contrived example of Quiz 20.2? It perfectly balances the length-1 jobs, leaving no good location for the length-5 job. If only the algorithm had considered the length-5 job first, all the other jobs would have fallen neatly into place. More generally, the second part of the intuition for Theorem 20.1 (page 47) and the final step in its proof (inequality (20.3)) both advocate for making the last job assigned to the most loaded machine (job j in (20.3)) as small as possible. This suggests the *longest processing time first (LPT)* algorithm (also proposed by Graham), which saves the smallest jobs for last.

LPT

Input/Output: as in the Graham algorithm (page 44).

---

sort the jobs from longest to shortest
run the Graham algorithm on the sorted jobs

The first step can be implemented in $O(n \log n)$ time (for n jobs) using, for example, the MergeSort algorithm. If the Graham algorithm is implemented with heaps (Problem 20.6), both of these steps run in near-linear time.[6]

Quiz 20.3

Suppose there are five machines, three jobs with length 5, two with length 6, two with length 7, two with length 8, and two with length 9. What is the makespan of the schedule output by the LPT algorithm, and what is the smallest-possible makespan of a schedule of these jobs?

a) 16 and 15

b) 17 and 15

c) 18 and 15

d) 19 and 15

(See Section 20.1.9 for the solution and discussion.)

Again, because the makespan minimization problem is NP-hard and the LPT algorithm runs in polynomial time, we fully expected such examples demonstrating that the latter is not always optimal. But does it provide a better insurance policy than the Graham algorithm?

Theorem 20.4 (LPT: Approximate Correctness) *The makespan of the schedule output by the LPT algorithm is always at most $\frac{3}{2} - \frac{1}{2m}$ times the minimum-possible makespan, where m denotes the number of machines.*

Intuitively, sorting the jobs reduces the possible damage caused by a single job—the difference between the largest and smallest machine loads—from M^* (the minimum-possible makespan) to $M^*/2$.

The keen reader may have noticed the daylight between the bad example in Quiz 20.3 (with a makespan blowup of $19/15 \approx 1.267$)

---

[6]The Graham algorithm is an example of an "online algorithm": It can be used even if the jobs materialize one by one and must be scheduled immediately. The LPT algorithm is not an online algorithm; it requires advance knowledge of all the jobs to sort them by length.

and the assurance of Theorem 20.4 (which, for $m = 5$, promises a blowup of at most $14/10 = 1.4$). With some additional arguments (outlined in Problem 20.7), the guarantee in Theorem 20.4 can be refined from $\frac{3}{2} - \frac{1}{2m}$ to $\frac{4}{3} - \frac{1}{3m}$. Consequently, the examples suggested by Quiz 20.3 are as bad as it gets for the LPT algorithm. And as with the Graham algorithm, you should expect the LPT algorithm to overdeliver for more realistic inputs.[7]

20.1.8   Proof of Theorem 20.4

The proof of Theorem 20.4 follows that of Theorem 20.1, with the improvement enabled by a variant of Lemma 20.2 that is useful when the jobs are sorted from longest to shortest.

Lemma 20.5 (Variant of Lower Bound #1) *If M^* denotes the minimum makespan of any schedule and j a job that is not among the m longest (breaking ties arbitrarily),*

$$M^* \geq 2\ell_j. \tag{20.4}$$

Proof: By the Pigeonhole Principle, every schedule must assign two of the longest $m + 1$ jobs to the same machine.[8] Therefore, the minimum-possible makespan is at least twice the length of the $(m + 1)$th-longest job; this is at least $2\ell_j$. $\mathcal{QED}$

Now on to:

Proof of Theorem 20.4: As in the final part of the proof of Theorem 20.1, let i denote a machine that has the largest load in the LPT algorithm's schedule and j the final job assigned to it (Figure 20.1(a)). Suppose at least one other job is assigned to i (prior to j); otherwise, there's nothing to prove.[9]

The algorithm assigns each of the first m jobs to a different machine (each empty at the time). Therefore, job j cannot be one of the first m

---

[7]There are more sophisticated algorithms with even better approximate correctness guarantees; these technically run in polynomial time but are impractically slow. If the makespan minimization problem comes up in your own work, the LPT algorithm is an excellent starting point.

[8]The *Pigeonhole Principle* is the self-evident fact that, no matter how you stuff $n + 1$ pigeons into n holes, there will be a hole with at least two pigeons.

[9]If j is the only job assigned to i, the algorithm's schedule has makespan ℓ_j and no other schedule can be better (by Lemma 20.2).

jobs. By the LPT greedy criterion, job j cannot be one of the m longest jobs. Lemma 20.5 then tells us that $\ell_j \leq M^*/2$, where M^* is the minimum-possible makespan. Plugging this improved bound (relative to Lemma 20.2) into the inequality (20.3) shows that the makespan M achieved by the LPT algorithm satisfies

$$M \leq \underbrace{\left(1 - \frac{1}{m}\right) \cdot \ell_j}_{\leq \left(1-\frac{1}{m}\right)\cdot(M^*/2)\text{ by (20.4)}} + \underbrace{\frac{1}{m}\sum_{h=1}^{n}\ell_h}_{\leq M^*\text{ by (20.2)}} \leq \left(\frac{3}{2} - \frac{1}{2m}\right) \cdot M^*.$$

$\mathcal{QED}$

20.1.9   Solutions to Quizzes 20.1–20.3

Solution to Quiz 20.1

Correct answer: (c). The machine loads are $2+2 = 4$ and $1+3 = 4$ in the first schedule and $2 + 3 = 5$ and $1 + 2 = 3$ in the second. Because the makespan is the largest machine load, these schedules have makespans of 4 and 5, respectively.

Solution to Quiz 20.2

Correct answer: (c). The Graham algorithm schedules the first twenty jobs evenly across the machines (with four length-1 jobs on each). No matter how it schedules the final length-5 job, it's stuck with a makespan of 9:

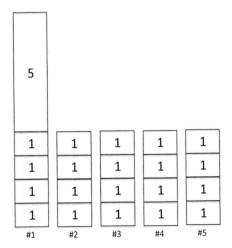

Meanwhile, reserving one machine for the big job and splitting the twenty small jobs evenly between the remaining four machines creates a perfectly balanced schedule, with makespan 5:

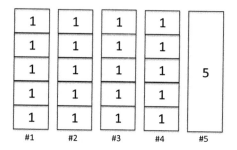

Solution to Quiz 20.3

Correct answer: (d). The optimal schedule is perfectly balanced, with the three length-5 jobs assigned to a common machine and every other machine receiving either a length-9 and length-6 job or a length-8 and length-7 job:

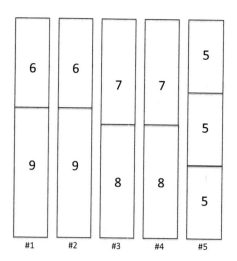

The makespan of this schedule is 15. Meanwhile, all machines already have load 14 when the time comes for the LPT algorithm to assign its final length-5 job, so it gets stuck with a final makespan of 19:

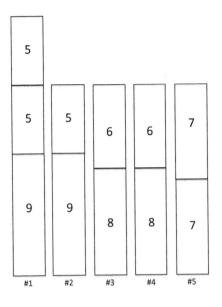

20.2   Maximum Coverage

Imagine you've been put in charge of assembling a team—maybe at your company to complete a project, or in your fantasy sports league to compete over a season. You can afford to hire only a limited number of people. Each potential team member has a combination of skills—perhaps corresponding to programming languages that they know, or positions on the field that they can play. You want a diverse team with as many different skills as possible. Whom should you pick?

20.2.1   Problem Definition

In the *maximum coverage* problem, the input comprises m subsets $A_1, A_2, \ldots, A_m$ of a ground set U, and a budget k. For example, in a team-hiring application, the ground set U corresponds to all possible skills that a team member could have, and each subset A_i corresponds to one potential team member, with the elements of the subset indicating the candidate's skills. The goal is to choose k of the subsets to maximize their *coverage*—the number of distinct ground set elements they contain. In a team-hiring problem, coverage corresponds to the number of distinct skills possessed by the team.

Problem: Maximum Coverage

Input: A collection $A_1, A_2, \ldots, A_m$ of subsets of a ground set U, and a positive integer k.

Output: A choice $K \subseteq \{1, 2, \ldots, m\}$ of k indices to maximize the coverage $f_{cov}(K)$ of the corresponding subsets, where:

$$f_{cov}(K) = \left| \cup_{i \in K} A_i \right|. \tag{20.5}$$

For example:

Quiz 20.4

Consider a ground set U with 16 elements and six subsets of it:

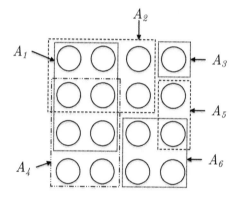

What is the largest coverage achieved by four of the subsets?

  a) 13

  b) 14

  c) 15

  d) 16

(See Section 20.2.8 for the solution and discussion.)

Maximum coverage problems are tricky because of overlaps between subsets. For example, some skills may be common (covered
by many subsets) and others rare (covered by few). An ideal subset
is large with few redundant elements—a team member blessed with
many unique skills.

20.2.2   Further Applications

Maximum coverage problems show up all the time, and not only in
team-hiring applications. For example, suppose you want to choose
locations for k new firehouses in a city to maximize the number of
residents who live within one mile of a firehouse. This is a maximum
coverage problem in which the ground set elements correspond to
residents, each subset corresponds to a possible firehouse location,
and the elements of a subset correspond to the residents who live
within one mile of that location.

For a more complex example, imagine you want to coax people to
show up at an event, such as a concert. You need to start setting up
for the event and have time to convince only k of your friends to come.
But whichever friends you recruit then bring along *their* friends, and
friends of friends, and so on. We can visualize this problem using a
directed graph, in which vertices correspond to people and an edge
directed from v to w signifies that w would follow v to the event
should v attend. For example, in the graph

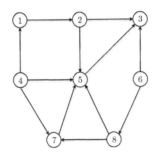

recruiting your friend 1 would ultimately trigger the attendance of four
people (1, 2, 3, and 5). Your friend 6 will show up only if recruited
directly, in which case four others (3, 5, 7, and 8) follow suit.

Maximizing event attendance is a maximum coverage problem.
Ground set elements correspond to people—equivalently, the vertices
of the graph. There is one subset per person, indicating who would

ultimately follow that person to the event—equivalently, the vertices reachable by a directed path from that vertex. The total attendance triggered by the recruitment of k people is then exactly the coverage achieved by the corresponding k subsets.

20.2.3   A Greedy Algorithm

The maximum coverage problem is NP-hard (see Problem 22.8). If we don't want to give up on speed, it's time to consider heuristic algorithms. Greedy algorithms, which myopically select subsets one by one, are again the obvious place to start.

The problem is easy to solve when you can pick only one subset ($k = 1$)—go with the biggest one. Suppose $k = 2$ and you have already committed to picking the biggest of the subsets, A. What should the second subset be? What matters now are the elements in a subset *not already covered by A*, so the sensible greedy criterion is to maximize the number of newly covered elements. Extending this idea to an arbitrary budget k leads to the following famous greedy algorithm for the maximum coverage problem, in which the coverage function f_{cov} is defined as in (20.5):[10]

GreedyCoverage

Input: subsets $A_1, A_2, \ldots, A_m$ of a ground set U and a positive integer k.
Output: a set $K \subseteq \{1, 2, \ldots, m\}$ of k indices.

---

1   $K := \emptyset$ // indices of chosen sets
2   **for** $j = 1$ to k **do**   // choose sets one by one
        // greedily increase coverage
        // (break ties arbitrarily)
3       $i^* := \operatorname{argmax}_{i=1}^{m} \left[f_{cov}(K \cup \{i\}) - f_{cov}(K) \right]$
4       $K := K \cup \{i^*\}$
5   **return** K

---

[10] First analyzed by Gérard P. Cornuéjols, Marshall L. Fisher, and George L. Nemhauser in the paper "Location of Bank Accounts to Optimize Float: An Analytic Study of Exact and Approximate Algorithms" (*Management Science*, 1977).

For simplicity, the argmax in line 3 examines all the subsets; equivalently, it could restrict attention to those not already chosen in previous iterations.

20.2.4   Bad Examples for the `GreedyCoverage` Algorithm

The `GreedyCoverage` algorithm is easy to implement in polynomial time.[11]   Because the maximum coverage problem is NP-hard, we should expect examples for which the algorithm outputs a suboptimal solution (as otherwise it would refute the $P \neq NP$ conjecture). Here's one:

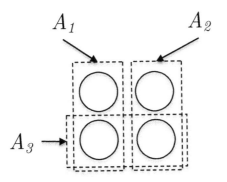

Suppose $k = 2$. The optimal solution is to pick the subsets A_1 and A_2 to cover all four elements. The greedy algorithm (with arbitrary tie-breaking) might well pick the subset A_3 in its first iteration, in which case it's stuck picking either A_1 or A_2 in the second iteration and covering only three of the four elements.

Are there still worse examples for the `GreedyCoverage` algorithm? At least for larger budgets k, the answer is yes.

Quiz 20.5

Consider the following ground set of 81 elements and five subsets of it:

---

[11]For example, compute the argmax in line 3 by exhaustive search through the m subsets, computing the additional coverage $f_{cov}(K \cup \{i\}) - f_{cov}(K)$ provided by a subset A_i using a single pass over A_i's elements. A straightforward implementation leads to a running time of $O(kms)$, where s denotes the maximum size of a subset (which is at most $|U|$).

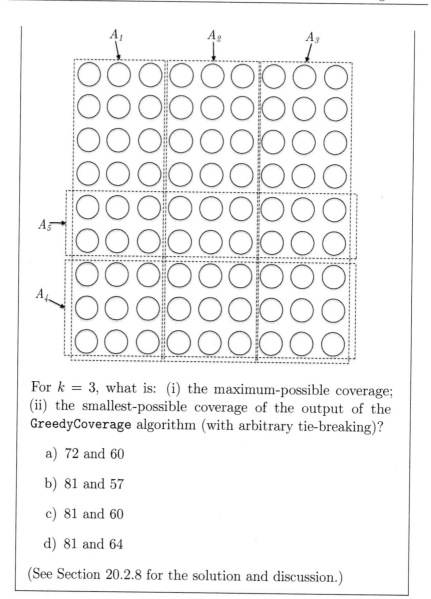

For $k = 3$, what is: (i) the maximum-possible coverage;
(ii) the smallest-possible coverage of the output of the
GreedyCoverage algorithm (with arbitrary tie-breaking)?

   a) 72 and 60

   b) 81 and 57

   c) 81 and 60

   d) 81 and 64

(See Section 20.2.8 for the solution and discussion.)

Thus, with $k = 2$, the GreedyCoverage algorithm might capture
only 75% of the elements that could be covered, and with $k = 3$,
it might fare as poorly as $\frac{57}{81} = \frac{19}{27} \approx 70.4\%$. How bad can things
get? Problem 20.8(a) asks you to extend this pattern to all positive
integers k, thereby showing:[12]

---

[12]The reliance on arbitrary tie-breaking is convenient but not essential to these
examples; see Problem 20.8(b).

Proposition 20.6 (Bad Examples for GreedyCoverage) *For every positive integer k, there is an instance of the maximum coverage problem in which:*

(a) There exist k subsets that cover the entire ground set.

(b) With arbitrary tie-breaking, the GreedyCoverage algorithm might cover only a $1 - (1 - \frac{1}{k})^k$ fraction of the elements.[13]

The easiest way to get a handle on a crazy expression with one variable is to plot it. Following this advice for the function $1 - (1 - \frac{1}{x})^x$, we see that it is decreasing but seems to approach an asymptote at roughly 63.2%:

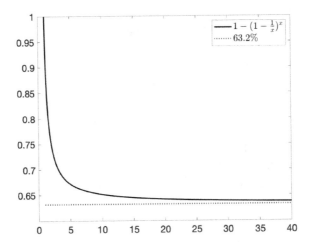

What's going on is that $1 - x$ is very well approximated by e^{-x} when x is close to 0 (as you should verify with a plot or a Taylor expansion of e^{-x}). Thus, the expression $1 - (1 - \frac{1}{k})^k$ tends to $1 - (e^{-1/k})^k = 1 - \frac{1}{e} \approx 0.632$ as k tends to infinity.[14]

20.2.5   Approximate Correctness

What's a weird number like $1 - \frac{1}{e}$ doing in proximity to the super-simple GreedyCoverage algorithm? Maybe it's an artifact of the

---

[13]Note that $1 - (1 - \frac{1}{k})^k$ equals $1 - (\frac{1}{2})^2 = \frac{3}{4}$ when $k = 2$ and $1 - (\frac{2}{3})^3 = \frac{19}{27}$ when $k = 3$.

[14]Here, $e = 2.718\ldots$ denotes Euler's number.

examples we cooked up in Quiz 20.5 and Proposition 20.6? Quite the opposite: The following approximate correctness guarantee proves that these are the worst examples for the GreedyCoverage algorithm, showing that it is inextricably tied to the numbers $1 - (1 - \frac{1}{k})^k$ and $1 - \frac{1}{e}$.[15,16]

Theorem 20.7 (GreedyCoverage: Approximate Correctness)
The coverage of the solution output by the GreedyCoverage algorithm is always at least a $1 - (1 - \frac{1}{k})^k$ fraction of the maximum-possible coverage, where k is the number of subsets chosen.

Thus, the GreedyCoverage algorithm is guaranteed to cover at least 75% as many elements as an optimal solution when $k = 2$, at least 70.4% as many when $k = 3$, and at least 63.2% as many no matter how large k is. As with Theorems 20.1 and 20.4, Theorem 20.7 is an insurance policy that limits the damage in a worst-case scenario; for more realistic inputs, the algorithm is likely to overdeliver and achieve significantly higher percentages.

20.2.6 A Key Lemma

To develop your intuition for Theorem 20.7, let's revisit the example in Quiz 20.5. Why didn't the GreedyCoverage algorithm come up with the optimal solution? In the first iteration, it had the option of picking any of the three subsets in the optimal solution (A_1, A_2, or A_3). Unfortunately, the algorithm was tricked by a fourth, equally large subset (A_4, covering 27 elements). In the second iteration, the algorithm again had the option of picking any of A_1, A_2, or A_3, but it was tricked by the subset A_5, which covered just as many new elements (18).

---

[15] It gets weirder: Assuming the P $\neq$ NP conjecture, *no* polynomial-time algorithm (greedy or otherwise) can guarantee a solution with coverage larger than a $1 - \frac{1}{e}$ fraction of the maximum possible as k grows large. (This is a difficult result, due to Uriel Feige in the paper "A Threshold of $\ln n$ for Approximating Set Cover" (*Journal of the ACM*, 1998).) This fact provides a strong theoretical justification for adopting the GreedyCoverage algorithm as a starting point when tackling the maximum coverage problem in practice. It also implies that the number $1 - \frac{1}{e}$ is intrinsic to the maximum coverage problem rather than an artifact of one particular algorithm.

[16] We won't see the number $1 - \frac{1}{e}$ again in this book, but it recurs mysteriously often in the analysis of algorithms.

In general, every miscue by the GreedyCoverage algorithm can be attributed to a subset that covers at least as many new elements as each of the k subsets in an optimal solution. But shouldn't this mean that the GreedyCoverage algorithm makes healthy progress in each iteration? This idea is formalized in the next lemma, which bounds from below the number of newly covered elements in each iteration as a function of the current coverage deficiency:

Lemma 20.8 (GreedyCoverage Makes Progress) *For each* $j \in \{1, 2, \ldots, k\}$, *let* C_j *denote the coverage achieved by the first* j *subsets chosen by the* GreedyCoverage *algorithm. For each such* j, *the* jth *subset chosen covers at least* $\frac{1}{k}(C^* - C_{j-1})$ *new elements, where* C^* *denotes the maximum-possible coverage by* k *subsets:*

$$C_j - C_{j-1} \geq \frac{1}{k}\left(C^* - C_{j-1}\right). \qquad (20.6)$$

Proof: Let K_{j-1} denote the indices of the first $j - 1$ subsets chosen by the GreedyCoverage algorithm and $C_{j-1} = f_{cov}(K_{j-1})$ their coverage. Consider any competing set $\widehat{K}$ of k indices, with corresponding coverage $\widehat{C} = f_{cov}(\widehat{K})$.

The most important inequality in the proof is:

$$\sum_{i \in \widehat{K}} \underbrace{[f_{cov}(K_{j-1} \cup \{i\}) - C_{j-1}]}_{\text{coverage increase from } A_i} \geq \underbrace{\widehat{C} - C_{j-1}}_{\text{current coverage gap}} . \qquad (20.7)$$

Why is it true? Let W denote the ground set elements covered by the subsets corresponding to $\widehat{K}$ but not those corresponding to K_{j-1} (Figure 20.2). On the one hand, the size of W is at least $\widehat{C} - C_{j-1}$, the right-hand side of (20.7). On the other, it is also no more than the left-hand side of (20.7): Each element of W contributes at least once to the sum—once per subset with index in $\widehat{K}$ that contains it. Therefore, the left-hand side of (20.7) is at least its right-hand side, with the size of W sandwiched between them.

Next, if the k numbers summed on the left-hand side of (20.7) were equal, each would be $\frac{1}{k}\sum_{i \in \widehat{K}}[f_{cov}(K_{j-1} \cup \{i\}) - C_{j-1}]$; otherwise, the largest of them would be even bigger:

$$\underbrace{\max_{i \in \widehat{K}}[f_{cov}(K_{j-1} \cup \{i\}) - C_{j-1}]}_{\text{biggest value}} \geq \underbrace{\frac{1}{k}\sum_{i \in \widehat{K}}[f_{cov}(K_{j-1} \cup \{i\}) - C_{j-1}]}_{\text{average value}}. \quad (20.8)$$

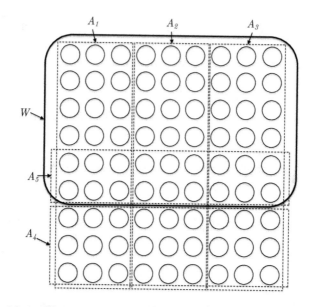

Figure 20.2: Illustration of the proof of Lemma 20.8 in the example from Quiz 20.5, with $j = 2$, $K_{j-1} = \{4\}$, $C_{j-1} = 27$, $\widehat{K} = \{1, 2, 3\}$, and $\widehat{C} = 81$. The set W is the $81 - 27 = 54$ elements covered by some subset with index in $\widehat{K}$ and no subset with index in K_{j-1}.

Now instantiate $\widehat{K}$ as the indices K^* of an optimal solution, with coverage $f_{cov}(K^*) = C^*$. Chaining together inequalities (20.7) and (20.8) shows that the GreedyCoverage algorithm has at least one good option (the best of the indices in the optimal solution K^*):

$$\underbrace{\max_{i \in K^*} [f_{cov}(K_{j-1} \cup \{i\}) - C_{j-1}]}_{\text{best of the optimal indices}} \geq \underbrace{\frac{1}{k} (C^* - C_{j-1})}_{\text{guaranteed progress}} .$$

The GreedyCoverage algorithm, due to its greedy criterion, selects an index that is at least this good, thereby increasing the coverage of its solution by at least $\frac{1}{k} (C^* - C_{j-1})$. $\mathcal{QED}$

20.2.7 Proof of Theorem 20.7

We can now prove Theorem 20.7 by iterating the recurrence (20.6) from Lemma 20.8 that bounds from below the progress made by the GreedyCoverage algorithm in each iteration. Continuing with the same notation, the goal is to compare the coverage C_k achieved by the algorithm's solution with the maximum-possible coverage C^*.

The anticipated term $1 - \frac{1}{k}$ enters the picture as soon as we apply Lemma 20.8 (first with $j = k$):

$$C_k \geq C_{k-1} + \frac{1}{k}(C^* - C_{k-1}) = \frac{C^*}{k} + \left(1 - \frac{1}{k}\right)C_{k-1}.$$

Applying it again (now with $j = k - 1$):

$$C_{k-1} \geq \frac{C^*}{k} + \left(1 - \frac{1}{k}\right)C_{k-2}.$$

Combining the two inequalities:

$$C_k \geq \frac{C^*}{k}\left(1 + \left(1 - \frac{1}{k}\right)\right) + \left(1 - \frac{1}{k}\right)^2 C_{k-2}.$$

Applying Lemma 20.8 a third time with $j = k-2$ and then substituting for C_{k-2}:

$$C_k \geq \frac{C^*}{k}\left(1 + \left(1 - \frac{1}{k}\right) + \left(1 - \frac{1}{k}\right)^2\right) + \left(1 - \frac{1}{k}\right)^3 C_{k-3}.$$

The pattern continues and, after k applications of the lemma (and using that $C_0 = 0$), we wind up with

$$C_k \geq \frac{C^*}{k}\underbrace{\left(1 + \left(1 - \frac{1}{k}\right) + \left(1 - \frac{1}{k}\right)^2 + \cdots + \left(1 - \frac{1}{k}\right)^{k-1}\right)}_{\text{geometric series}}.$$

Inside the parentheses is an old friend, a geometric series. In general, for $r \neq 1$, there is a useful closed-form formula for a geometric series:[17]

$$1 + r + r^2 + \cdots + r^\ell = \frac{1 - r^{\ell+1}}{1 - r}. \tag{20.9}$$

Invoking this formula with $r = 1 - \frac{1}{k}$ and $\ell = k - 1$, our lower bound on C_k transforms into

$$C_k \geq \frac{C^*}{k}\left(\frac{1 - (1 - \frac{1}{k})^k}{1 - (1 - \frac{1}{k})}\right) = C^*\left(1 - \left(1 - \frac{1}{k}\right)^k\right),$$

fulfilling the promise made by Theorem 20.7. $\mathcal{QED}$

---

[17] To verify this identity, multiply both sides by $1-r$: $(1-r)(1+r+r^2+\cdots+r^\ell) = 1 - r + r - r^2 + r^2 - r^3 + r^3 - \cdots - r^{\ell+1} = 1 - r^{\ell+1}$.

20.2.8   Solutions to Quizzes 20.4–20.5

Solution to Quiz 20.4

Correct answer: (c). There are two ways to cover 15 of the 16 elements, by choosing A_2, A_4, A_6, and either A_3 or A_5. The large subset A_1 does not participate in any optimal solution because it is largely redundant with the other subsets.

Solution to Quiz 20.5

Correct answer: (b). The optimal solution picks the subsets A_1, A_2, and A_3 and covers all 81 elements. One possible execution of the GreedyCoverage algorithm picks A_4 in its first iteration, breaking a four-way tie with A_1, A_2, A_3; A_5 in its second iteration, again breaking a four-way tie with A_1, A_2, A_3; and, finally, A_1. This solution has coverage $27 + 18 + 12 = 57$.

*20.3   Influence Maximization

The GreedyCoverage algorithm from Section 20.2 was originally motivated by old-school applications like choosing locations for new factories. In the 21st century, generalizations of this algorithm have found new applications in many fields of computer science. This section describes a representative example in social network analysis.[18]

20.3.1   Cascades in Social Networks

For our purposes, a *social network* is a directed graph $G = (V, E)$ in which the vertices correspond to people and a directed edge (v, w) signifies that v "influences" w. For example, perhaps w follows v in an online social network such as Twitter or Instagram.

A *cascade model* posits how information (such as a news article or meme) travels through a social network. Here's a simple one, parameterized by a directed graph $G = (V, E)$, an activation probability $p \in [0, 1]$, and a subset $S \subseteq V$ of *seed* vertices:[19]

---

[18]Starred sections like this one are the more difficult sections; they can be skipped on a first reading.

[19]To brush up on basic discrete probability, see Appendix B of *Part 1* or the resources at www.algorithmsilluminated.org.

A Simple Cascade Model

Initially, every seed vertex is "active" and all other vertices are "inactive." All edges are initially "unflipped."

While there is some active vertex v and unflipped outgoing edge (v, w):

- Flip a biased coin that comes up "heads" with probability p.

- If the coin comes up "heads," update the status of edge (v, w) to "active." If w is inactive, update its status to "active."

- If the coin comes up "tails," update the status of edge (v, w) to "inactive."

Once a vertex is activated (say, due to reading an article or seeing a movie), it never becomes inactive. A vertex can have multiple activation opportunities—one for each of its activated influencers. For example, maybe the first two recommendations from friends for a new movie don't register, but the third triggers you to go see it.

20.3.2   Example

In the graph

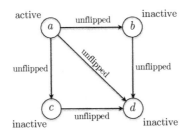

the vertex a is a seed and initially active; the rest are initially inactive. Each of the outgoing edges (a, b), (a, c), and (a, d) has a probability of p of activating the other endpoint of the edge. Suppose the coin associated with edge (a, b) comes up "heads" and the other two come up "tails." The new picture is:

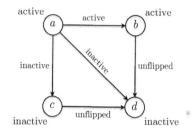

At this point, there is no hope of activating vertex c. There remains a probability of p that vertex d is activated via the unflipped edge (b, d); if this event occurs, the final state is:

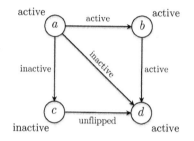

For closure and convenience, we can optionally add a postprocessing step that flips the coins of any remaining unflipped edges and updates their statuses accordingly (while leaving all vertices' statuses unchanged). In our running example, the final result might be:

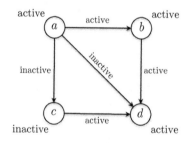

In general, with or without the postprocessing step, the vertices that wind up activated at the end of the process are precisely those reachable from a seed vertex by a directed path of activated edges.

20.3.3   The Influence Maximization Problem

In the *influence maximization* problem, the goal is to choose a limited number of seed vertices in a social network to maximize the spread of information, meaning the number of vertices that are eventually activated according to our cascade model.[20] This number is a random variable, depending on the outcomes of the coin flips in the cascade model, and we focus on its expectation.[21] Formally, let $X(S)$ denote the (random) set of vertices that are eventually activated when the vertices S are chosen as seeds, and define the *influence* of S as

$$f_{inf}(S) = \mathbf{E}[|X(S)|], \qquad (20.10)$$

where the expectation is over the random coin flips in the cascade model. The influence of a set depends on both the graph and on the activation probability, with more edges or a higher probability resulting in a bigger influence.

Problem: Influence Maximization

Input: A directed graph $G = (V, E)$, a probability p, and a positive integer k.

Output: A choice $S \subseteq V$ of k vertices with the maximum-possible influence $f_{inf}(S)$ in the cascade model with activation probability p.

For example, if you're giving away k promotional copies of a product and want to choose the recipients to maximize its eventual adoption, you're facing an influence maximization problem.

Problem 20.9 asks you to show that the maximum coverage problem can be viewed as a special case of the influence maximization problem. Because the special case is NP-hard (Problem 22.8), so is the more general problem. Could there be a fast and approximately correct heuristic algorithm for the influence maximization problem?

---

[20] For much more on the influence maximization problem and its many variations, check out the paper "Maximizing the Spread of Influence Through a Social Network," by David Kempe, Jon Kleinberg, and Éva Tardos (*Theory of Computing*, 2015).

[21] The *expectation* $\mathbf{E}[Y]$ of a random variable Y is its average value, weighted by the appropriate probabilities. For example, if Y can take on the values $\{0, 1, 2, \ldots, n\}$, then $\mathbf{E}[Y] = \sum_{i=0}^{n} i \cdot \mathbf{Pr}[Y = i]$.

20.3.4   A Greedy Algorithm

The influence maximization problem resembles the maximum coverage problem, with vertices playing the role of subsets and with influence (20.10) playing the role of coverage (20.5). The GreedyCoverage algorithm for the latter problem translates easily to the former, swapping in the new definition (20.10) of the objective function.

GreedyInfluence

Input: directed graph $G = (V, E)$,
probability $p \in [0, 1]$, and positive integer k.
Output: a set $S \subseteq V$ of k vertices.

---

1 $S := \emptyset$ // chosen vertices
2 **for** $j = 1$ to k **do**   // choose vertices one by one
        // greedily increase influence
        // (break ties arbitrarily)
3        $v^* := \text{argmax}_{v \in V} \left[f_{inf}(S \cup \{v\}) - f_{inf}(S) \right]$
4        $S := S \cup \{v^*\}$
5 **return** S

Quiz 20.6

What is the running time of a straightforward implementation of the GreedyInfluence algorithm on graphs with n vertices and m edges? (Choose the strongest true statement.)

a) $O(knm)$

b) $O(knm^2)$

c) $O(knm2^m)$

d) Unclear

(See Section 20.3.8 for the solution and discussion.)

20.3.5   Approximate Correctness

Happily, our greedy heuristic algorithm remains equally approximately correct for the influence maximization problem. Because the maximum coverage problem is a special case of the influence maximization problem (Problem 20.9), this is the best-case scenario—an equally strong approximate correctness guarantee, but for a more general problem.

Theorem 20.9 (GreedyInfluence: Approximate Correctness)
The influence of the solution output by the GreedyInfluence algorithm is always at least a $1 - (1 - \frac{1}{k})^k$ fraction of the maximum-possible influence, where k is the number of vertices chosen.

The key insight in the proof of Theorem 20.9 is to recognize the influence function (20.10) as a weighted average of coverage functions (20.5). Each of these coverage functions corresponds to the event attendance application (Section 20.2.2) with a subgraph of the social network (comprising the activated edges). We can then check that the proof of Theorem 20.7 in Sections 20.2.6 and 20.2.7 for coverage functions can be extended to weighted averages of coverage functions.

20.3.6   Influence Is a Weighted Sum of Coverage Functions

More formally, fix a directed graph $G = (V, E)$, an activation probability $p \in [0, 1]$, and a positive integer k. For convenience, we include the postprocessing step in the cascade model (see Section 20.3.2) so that every edge ends up either active or inactive. The vertices $X(S)$ activated by the seeds S are precisely those reachable from a vertex in S by a directed path of activated edges.

As a thought experiment, imagine we had telepathy and knew in advance the edges $H \subseteq E$ that would be activated—in effect, tossing all edges' coins up front rather than on a need-to-know basis. Then, the influence maximization problem would boil down to a maximum coverage problem. The ground set would be the vertices V and there would be one subset per vertex, with the subset $A_{v,H}$ containing the vertices that are reachable from v by a directed path in the subgraph (V, H) of activated edges. For example, if the graph and edge statuses are:

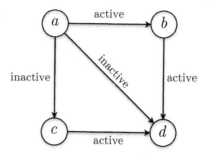

then $A_{a,H} = \{a, b, d\}$, $A_{b,H} = \{b, d\}$, $A_{c,H} = \{c, d\}$, and $A_{d,H} = \{d\}$. The influence of a set S of seeds (given the activated edges H) is then the coverage of the corresponding subsets:

$$f_H(S) := |\cup_{v \in S} A_{v,H}|. \tag{20.11}$$

Of course, we don't have advance knowledge of the subset of activated edges, and each subset $H \subseteq E$ occurs with some positive probability p_H.[22] But because influence is defined as an expectation, we can express it as a weighted average of coverage functions, with weights equal to probabilities:[23]

Lemma 20.10 (Influence = Average of Coverage Functions)
For each subset $H \subseteq E$ of edges, let f_H denote the coverage function defined in (20.11) and p_H the probability that the subset of edges activated in the cascade model is precisely H. For every subset $S \subseteq V$ of vertices,

$$f_{inf}(S) = \mathbf{E}_H[f_H(S)] = \sum_{H \subseteq E} p_H \cdot f_H(S). \tag{20.12}$$

20.3.7   Proof of Theorem 20.9

We can declare victory after proving an analog of Lemma 20.8 showing that the `GreedyInfluence` algorithm makes healthy progress in each iteration. Theorem 20.9 then follows from the exact same algebra that we used to prove Theorem 20.7 in Section 20.2.7.

---

[22]Not that we'll need it, but the formula is $p_H = p^{|H|}(1-p)^{|E|-|H|}$.

[23]For the rigor-obsessed: We're using the law of total expectation to write the expectation in (20.10) as a probability-weighted average of conditional expectations, where the conditioning is on the activated edges H.

Lemma 20.11 (GreedyInfluence Makes Progress) *For each* $j \in \{1, 2, \ldots, k\}$, *let* I_j *denote the influence achieved by the first* j *vertices chosen by the* GreedyInfluence *algorithm. For each such* j, *the* jth *vertex chosen increases the influence by at least* $\frac{1}{k}(I^* - I_{j-1})$, *where* I^* *denotes the maximum-possible influence of* k *vertices:*

$$I_j - I_{j-1} \geq \frac{1}{k}\left(I^* - I_{j-1}\right).$$

Proof: Let S_{j-1} denote the first $j-1$ vertices chosen by the GreedyInfluence algorithm and $I_{j-1} = f_{inf}(S_{j-1})$ their influence. Let S^* denote a set of k vertices with the maximum-possible influence I^*.

Next, consider an arbitrary subset $H \subseteq E$ of edges and the corresponding coverage function f_H defined in (20.11). Piggybacking on our hard work for coverage functions, we can translate the key inequality (20.7) in the analysis of the GreedyCoverage algorithm to the inequality

$$\sum_{v \in S^*} \underbrace{[f_H(S_{j-1} \cup \{v\}) - f_H(S_{j-1})]}_{\text{coverage increase from } v \text{ (under } f_H)} \geq \underbrace{f_H(S^*) - f_H(S_{j-1})}_{\text{coverage gap (under } f_H)}; \quad (20.13)$$

here S_{j-1} and S^* are playing the roles of K_{j-1} and $\widehat{K}$, and $f_H(S_{j-1})$ and $f_H(S^*)$ correspond to C_{j-1} and $\widehat{C}$.

From Lemma 20.10, we know that influence (f_{inf}) is a weighted average of coverage functions (the f_H's). We have one inequality of the form (20.13) for each subset $H \subseteq E$ of edges; for shorthand, denote its left- and right-hand sides by L_H and R_H, respectively. The idea is to examine the analogous weighted average of these 2^m inequalities (where m denotes $|E|$).

Because multiplying both sides of an inequality by the same nonnegative number (such as a probability p_H) preserves it,

$$p_H \cdot L_H \geq p_H \cdot R_H$$

for every $H \subseteq E$. Because all 2^m inequalities go in the same direction, they add up to a combined inequality:

$$\sum_{H \subseteq E} p_H \cdot L_H \geq \sum_{H \subseteq E} p_H \cdot R_H. \quad (20.14)$$

Unpacking the right-hand side of (20.14) and using the expanded formula for f_{inf} in (20.12), we obtain

$$\sum_{H \subseteq E} p_H \left(f_H(S^*) - f_H(S_{j-1})\right) = \sum_{H \subseteq E} p_H \cdot f_H(S^*) - \sum_{H \subseteq E} p_H \cdot f_H(S_{j-1})$$

$$= \underbrace{f_{inf}(S^*) - f_{inf}(S_{j-1})}_{\text{right-hand side of (20.14)}}.$$

The left-hand side of (20.14), after the same maneuvers, becomes

$$\underbrace{\sum_{v \in S^*} [f_{inf}(S_{j-1} \cup \{v\}) - f_{inf}(S_{j-1})]}_{\text{left-hand side of (20.14)}}.$$

Thus, inequality (20.14) translates to an analog of the key inequality (20.7) in the proof of Lemma 20.8:

$$\sum_{v \in S^*} [\underbrace{f_{inf}(S_{j-1} \cup \{v\}) - \underbrace{f_{inf}(S_{j-1})}_{=I_{j-1}}] \geq \underbrace{f_{inf}(S^*)}_{=I^*} - \underbrace{f_{inf}(S_{j-1})}_{=I_{j-1}}. \quad (20.15)$$

The biggest of the k terms in the sum on the left-hand side is at least the average value (as in (20.8)), so the GreedyInfluence algorithm always has at least one good option (the best of the vertices in the optimal solution S^*):

$$\max_{v \in S^*} [f_{inf}(S_{j-1} \cup \{v\}) - I_{j-1}] \geq \frac{1}{k} (I^* - I_{j-1}).$$

The GreedyInfluence algorithm, due to its greedy criterion, selects a vertex that is at least this good, thereby increasing the influence of its solution by at least $\frac{1}{k}(I^* - I_{j-1})$. $\mathcal{QED}$

20.3.8   Solution to Quiz 20.6

Correct answers: (c),(d). There are k iterations of the main loop, each of which involves computing an argmax over the n vertices. The running time of a straightforward implementation is, therefore, $O(kn)$ times the number of operations required to compute the influence $f_{inf}(S)$ of a subset S. And how many operations is this? Unlike for the coverage objective function f_{cov}, the answer is not obvious because of the pesky expectation in (20.10). (In this sense, answer (d) is correct.) Computing this expectation naively—computing $|X(S)|$ via breadth- or depth-first search in $O(m)$ time for each of the 2^m possible outcomes of the coin flips, and averaging the results—leads to the running time bound in (c).

So is the GreedyInfluence algorithm useless in practice? Not at all. The influence $f_{inf}(S)$ of a subset S may be difficult to compute to arbitrary precision, but it is easy to estimate accurately using random sampling. In other words, given a subset S, go ahead and flip all the coins in the cascade model and see how many vertices end up activated with seed set S. After repeating this experiment many times, the average number of activated vertices will almost always be a good estimate of $f_{inf}(S)$.

20.4 The 2-OPT Heuristic Algorithm for the TSP

NP-hardness is always a drag, but at least for the NP-hard problems in Sections 20.1–20.3 (makespan minimization, maximum coverage, and influence maximization), there are fast algorithms with good approximate correctness guarantees. Alas, for many other NP-hard problems, including the TSP, such an algorithm would refute the $P \neq NP$ conjecture (see Problem 22.12). If you insist on an efficient algorithm for such a problem, the best-case scenario is a heuristic algorithm that, despite having no insurance policy, works well on many of the problem instances that arise in your application. *Local search*, along with its many variants, is one of the most powerful and flexible paradigms for devising algorithms of this type.

20.4.1 Tackling the TSP

I'm not going to tell you what local search is in general just yet. Instead, we'll devise from scratch a heuristic algorithm for the traveling salesman problem (TSP), which will force us to develop several new ideas. Then, in Section 20.5, we'll zoom out and identify the ingredients of our solution that exemplify the general principles of local search. Armed with a template for developing local search algorithms and an example instantiation, you'll be well-positioned to apply the technique in your own work.

In the TSP (Section 19.1.2), the input is a complete graph $G = (V, E)$ with real-valued edge costs, and the goal is to compute a tour— a cycle visiting every vertex exactly once—with the minimum-possible sum of edge costs. The TSP is NP-hard (see Section 22.7); if speed is mission-critical, the only option is to resort to heuristic algorithms (assuming, as usual, that the $P \neq NP$ conjecture is true).

To get a feel for the TSP, let's start with the first greedy algorithm that you might think of, along the lines of Prim's minimum spanning tree algorithm.

Quiz 20.7

The *nearest neighbor* heuristic algorithm is a greedy algorithm for the TSP which, given a complete graph with real-valued edge costs, works as follows:

1. Begin a tour at an arbitrary vertex a.

2. Repeat until all vertices have been visited:

   a) If the current vertex is v, proceed to the closest unvisited vertex (a vertex w minimizing c_{vw}).

3. Return to the starting vertex.

In the following example, what is the cost of the tour constructed by the nearest neighbor algorithm, and what is the minimum cost of a tour?

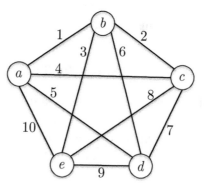

a) 23 and 29

---

b) 24 and 29

c) 25 and 29

d) 24 and 30

(See Section 20.4.6 for the solution and discussion.)

Quiz 20.7 shows that the nearest neighbor heuristic algorithm does not always construct a minimum-cost tour—hardly surprising, given that the TSP is NP-hard and the algorithm runs in polynomial time. More disturbingly, the greedily constructed tour remains the same even if we change the cost of the final hop (a, e) to a huge number. Unlike our greedy heuristic algorithms in Sections 20.1–20.3, the nearest neighbor algorithm can produce solutions that are worse than an optimal solution by an arbitrarily large factor. More sophisticated greedy algorithms can overcome this particular bad example, but all ultimately suffer an equally disappointing fate in more complicated TSP instances.

20.4.2   Improving a Tour with 2-Changes

Who says we have to give up as soon as we've constructed an initial tour? If there's a way to greedily tweak a tour to make it better, why not do it? What's the minimal modification that could transform one tour into a better one?

Quiz 20.8

In a TSP instance with n vertices, what is the maximum number of edges that two distinct tours can share?

a) $\log_2 n$

b) $n/2$

c) $n - 2$

d) $n - 1$

(See Section 20.4.6 for the solution and discussion.)

Quiz 20.8 suggests exploring the landscape of tours by swapping out one pair of edges for another:

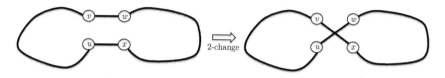

This type of swap is called a *2-change*.

2-Change

1. Given a tour T, remove two edges (v, w), (u, x) of T that do not share an endpoint.

2. Add either the edges (v, x) and (u, w) or the edges (u, v) and (w, x), whichever pair leads to a new tour T'.

The first step chooses two edges with four distinct endpoints.[24] There are three different ways to pair up these four vertices, and exactly one of them creates a new tour (as in the preceding figure). Beyond this and their original pairing, the third pairing creates two disjoint cycles rather than a feasible tour:

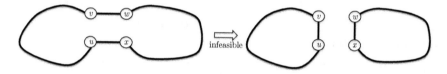

A 2-change can create a tour that is better or worse than the original one. If the newly swapped-in edges are (u, w) and (v, x):

$$\text{decrease in tour cost} = \underbrace{(c_{vw} + c_{ux})}_{\text{edges removed}} - \underbrace{(c_{uw} + c_{vx})}_{\text{edges added}}. \qquad (20.16)$$

If the decrease in (20.16) is positive—if the benefit $c_{vw} + c_{ux}$ of removing the old edges outweighs the cost $c_{uw} + c_{vx}$ of adding the new ones—the 2-change produces a lower-cost tour and is called *improving*.

---

[24]Removing two edges with a shared endpoint is pointless; the only way to get back a feasible tour would be to put them right back in.

For example, starting from the greedily constructed tour in Quiz 20.7, there are five candidate 2-changes, three of which are improving:[25]

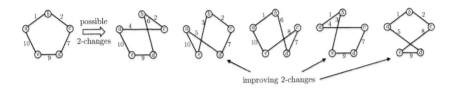

20.4.3   The 2-OPT Algorithm

The *2-OPT* algorithm for the TSP constructs an initial tour (for example, with the nearest neighbor algorithm) and performs improving 2-changes until none remain. In the following pseudocode, 2Change is a subroutine that takes as input a tour and two of its edges (with distinct endpoints) and returns as output the tour produced by the corresponding 2-change (as described on page 78).

2OPT

Input: complete graph $G = (V, E)$ and cost c_e for each edge $e \in E$.

Output: a traveling salesman tour.

---

1   $T :=$ initial tour   // perhaps greedily constructed
2   **while** improving 2-change $(v, w), (u, x) \in T$ exists **do**
3      $T := 2\text{Change}(T, (v, w), (u, x))$
4   return T

For example, starting from the tour constructed by the nearest neighbor algorithm in Quiz 20.7, the first iteration of the 2OPT algorithm might replace the edges (a, b) and (d, e) with the edges (a, d) and (b, e):

---

[25]In general, with $n \geq 4$ vertices, there are always $n(n - 3)/2$ candidate 2-changes: Choosing one of the n tour edges followed by one of the $n - 3$ tour edges with different endpoints counts every 2-change in exactly two different ways.

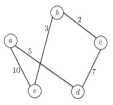

thereby lowering the tour cost from 29 to 27. From here, there are again five 2-changes to consider:

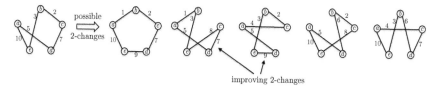

improving 2-changes

If the second iteration of the algorithm executes the first of the two improving 2-changes, replacing the edges (a, e) and (b, c) with the edges (a, b) and (c, e), the tour cost decreases further from 27 to 24. At this point, there are no improving 2-changes (one leaves the tour cost unchanged and four increase it), and the algorithm halts:

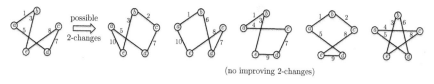

(no improving 2-changes)

20.4.4　Running Time

Does the 2OPT algorithm even halt, or could it loop forever? There are an awful lot of traveling salesman tours (Quiz 19.1), but still only finitely many. Every iteration of the 2OPT algorithm produces a tour with cost strictly smaller than the previous one, so there's no worry about the same tour showing up in two different iterations. Even in the doomsday scenario in which the algorithm searches through every possible tour, it halts within a finite amount of time.

The running time of the algorithm is governed by the number of iterations of the main while loop, times the number of operations performed per iteration. With n vertices, there are $O(n^2)$ different 2-changes to check in each iteration, leading to a per-iteration time bound of $O(n^2)$ (Problem 20.13). What about the number of iterations?

The bad news is that, in pathological examples, the 2OPT algorithm might perform an exponential (in n) number of iterations before halting. The good news is twofold. First, on more realistic inputs, the 2OPT algorithm almost always halts in a reasonable number of iterations (typically subquadratic in n). Second, because the algorithm maintains a feasible tour throughout its execution, it can be interrupted at any time.[26] You can decide in advance how long you're willing to run the algorithm (one minute, one hour, one day, etc.) and, when time expires, use the last (and best) solution that the algorithm found.

20.4.5   Solution Quality

The 2OPT algorithm can only improve upon its initial tour, but there's no guarantee that it will find an optimal solution. Already for the example in Quiz 20.7, we saw in Section 20.4.3 that the algorithm might return a tour with cost 24 instead of the optimal tour (which has cost 23). Could there be worse examples? How much worse?

The bad news is that more complicated and contrived examples show that the tour returned by the 2OPT algorithm can cost more than an optimal tour by an arbitrarily large factor. In other words, the algorithm does not have an approximate correctness guarantee akin to those in Sections 20.1–20.3. The good news is that, for the instances of the TSP that arise in practice, variants of the 2OPT algorithm routinely find tours with total cost not much more than the minimum possible. To tackle the TSP in practice on large inputs (with n in the thousands or more), the 2OPT algorithm, augmented with some of the bells and whistles covered in Section 20.5, is an excellent starting point.

20.4.6   Solutions to Quizzes 20.7–20.8

Solution to Quiz 20.7

Correct answer: (a). The nearest neighbor algorithm starts at a and greedily travels to b and then c. At this point, the tour must proceed to either d or e (as no vertex can be visited twice), with d

---

[26]Algorithms that are interruptible in this sense are sometimes called *anytime algorithms*.

slightly preferred (cost 7 instead of 8). Once at d, the remaining hops of the tour are forced: There is no choice but to travel to e (at a cost of 9) and then return to a (at a cost of 10). The total cost of this tour is 29 (see the left figure below). Meanwhile, the minimum total cost of a tour is 23 (see the right figure below).

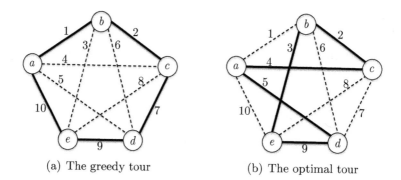

(a) The greedy tour   (b) The optimal tour

Solution to Quiz 20.8

Correct answer: (c). Because any $n - 1$ edges of a tour uniquely determine its final edge, distinct tours cannot share $n - 1$ edges. Distinct tours can share $n - 2$ edges, however:

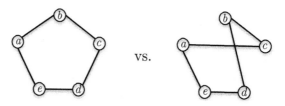

vs.

20.5 Principles of Local Search

A *local search algorithm* explores a space of feasible solutions via "local moves" that successively improve an objective function. The 2OPT algorithm for the TSP, in which local moves correspond to 2-changes, is a canonical example. This section zooms out and isolates the essential ingredients of the local search algorithm design paradigm, together with the key modeling and algorithmic decisions needed to apply it.

20.5.1   The Meta-Graph of Feasible Solutions

For a TSP instance $G = (V, E)$ with real-valued edge costs, the 2OPT algorithm can be visualized as a greedy walk through a "meta-graph" $H = (X, F)$ of feasible solutions (shown in Figure 20.3 for the running example from Quiz 20.7). The meta-graph H has one vertex $x \in X$ for each tour of G, labeled with the tour's total cost. It also has one edge $(x, y) \in F$ for each pair x, y of tours that differ in exactly two edges of G. In other words, the meta-graph edges correspond to the possible 2-changes in the TSP instance.

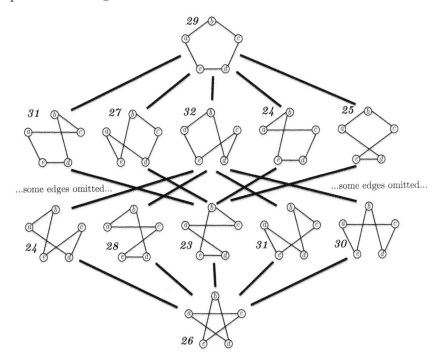

Figure 20.3: The meta-graph of feasible solutions corresponding to the TSP instance in Quiz 20.7. Vertices of the meta-graph correspond to tours, and two tours are connected by an edge of the meta-graph if and only if they differ in exactly two edges. The top tour is adjacent in the meta-graph to the five tours in the second row, and similarly for the bottom tour and the tours in the third row. Each tour in the second row is adjacent to each tour in the third row, save the one in its own column. (To avoid clutter, some of these meta-graph edges are omitted in the figure.) Each tour is labeled with its total cost.

Quiz 20.9

For a TSP instance with $n \geq 4$ vertices, how many vertices and edges does the corresponding meta-graph have?[27]

a) $\frac{1}{2}(n-1)!$ and $\frac{n!(n-3)}{8}$

b) $\frac{1}{2}(n-1)!$ and $\frac{n!(n-3)}{4}$

c) $(n-1)!$ and $\frac{n!(n-3)}{4}$

d) $(n-1)!$ and $\frac{n!(n-3)}{2}$

(See Section 20.5.8 for the solution and discussion.)

We can view the 20PT algorithm as starting at some meta-graph vertex (for example, the output of the nearest neighbor algorithm) and repeatedly traversing meta-graph edges to visit a sequence of tours with successively smaller costs. The algorithm halts when it reaches a meta-graph vertex with cost no larger than any of its neighbors in the meta-graph. The example 20PT trajectory in Section 20.4.3 begins at the top tour in Figure 20.3 before proceeding to the second tour in the second row and then halting at the first tour in the third row.

20.5.2   The Local Search Algorithm Design Paradigm

Most local search algorithms can be similarly visualized as a greedy walk in a meta-graph of feasible solutions.[28] Such algorithms differ only in the choice of meta-graph and the details of the exploration strategy.[29]

---

[27] Don't worry about the meta-graph being REALLY BIG; it exists only in our minds, never to be written out explicitly.

[28] We can even add a third dimension to the visualization, with the "height" of a meta-graph vertex specified by its objective function value. This image explains why local search is sometimes called *hill climbing*.

[29] One variant is *gradient descent*, an ancient local search algorithm for continuous (as opposed to discrete) optimization that is central to modern machine learning. The simplest version of gradient descent is a heuristic algorithm for minimizing a differentiable objective function over all points in Euclidean space, with improving local moves corresponding to small steps in the direction of steepest descent (that is, of the negative gradient) from the current point.

The Local Search Paradigm

1. Define your feasible solutions (equivalently, the vertices of the meta-graph).

2. Define your objective function (the numerical labels of the meta-graph vertices) and whether the goal is to maximize or minimize it.

3. Define your allowable local moves (the edges of the meta-graph).

4. Decide how to choose an initial feasible solution (a starting meta-graph vertex).

5. Decide how to choose among multiple improving local moves (the possible next steps in the meta-graph).

6. Perform local search: Starting from the initial feasible solution, iteratively improve the objective function value via local moves until reaching a *local optimum*— a feasible solution from which no improving local move is possible.

The pseudocode of a generic local search algorithm closely resembles that of the 2OPT algorithm. (MakeMove takes as input a feasible solution and the description of a local move, and returns the corresponding neighboring solution.)

GenericLocalSearch

$S :=$ initial solution   // as specified in step 4
while improving local move L exists **do**
    $S :=$ MakeMove(S, L) // as specified in step 5
return S // return the local optimum found

The first three steps in the local search paradigm are modeling decisions and the next two are algorithmic decisions. Let's examine each step in more detail.

20.5.3   Three Modeling Decisions

The first two steps of the local search paradigm define the problem.

Step 1: Define your feasible solutions. In the TSP, the feasible solutions of an n-vertex instance are the $\frac{1}{2}(n-1)!$ tours. In the makespan minimization problem (Section 20.1.1), the feasible solutions of an instance with m machines and n jobs are the m^n different ways of assigning jobs to machines. In the maximum coverage problem (Section 20.2.1), in an instance with m subsets and parameter k, the feasible solutions are the $\binom{m}{k}$ different ways of choosing k of the subsets.

Step 2: Define your objective function. This step is even more straightforward in our running examples. In the TSP, the objective function (to minimize) is the total cost of a tour. In the makespan minimization and maximum coverage problems, the objective functions (to minimize and maximize, respectively) are, naturally, the makespan of a schedule and the coverage of a collection of k subsets.

Steps 1 and 2 define the *global optima* of an instance—the feasible solutions with the best-possible objective function value (like the tour with cost 23 in Figure 20.3). Step 3 completes the definition of the meta-graph by specifying its edges—the permitted local moves from one feasible solution to another.

Step 3: Define your allowable local moves. In the 2OPT algorithm for the TSP, local moves correspond to 2-changes; in an instance with n vertices, the *neighborhood size*—the number of local moves available from each solution—is $n(n-3)/2$. What if you wanted to apply the local search paradigm to the makespan minimization or maximum coverage problem? For the former, the simplest definition of a local move is the reassignment of a single job to a different machine. The neighborhood size is then $n(m-1)$, where m and n denote the number of machines and jobs, respectively. For the maximum coverage problem, the simplest type of local move swaps out one of the k subsets in the current solution for a different one. With m subsets, the neighborhood size is then $k(m-k)$, with k choices to swap out and $m-k$ to swap in.

The meta-graph of feasible solutions is fully specified after steps 1–3 and so are the *local optima*—the feasible solutions from which there is no improving local move or, equivalently, the meta-graph vertices with objective function value at least as good as all their neighbors. For example, in Figure 20.3, the two local optima are the first and third tours in the third row. (The latter is also a global optimum, while the former is not.) For the makespan minimization problem, with local moves corresponding to single job reassignments, the schedule produced by the LPT algorithm in Quiz 20.3 is a local minimum, while that produced by the Graham algorithm in Quiz 20.2 is not (as you should check). For the maximum coverage problem, with local moves corresponding to subset swaps, the output of the GreedyCoverage algorithm is not a local maximum in either of the examples in Section 20.2.4 (as you should check).[30]

Example: The 3-Change Neighborhood for the TSP

Steps 1 and 2 do not uniquely determine the decision in step 3, as multiple definitions of a "local move" might make sense for a given computational problem. For example, in the TSP, who says we can only take out and put back in two edges at a time? Why not three, or more?

A *3-change* is an operation that replaces three edges of a traveling salesman tour with three different edges in a way that produces a new tour:[31]

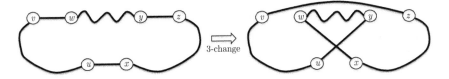

The 3OPT algorithm is the generalization of the 2OPT algorithm (Section 20.4.3) that, in each iteration of the main while loop, performs a 2-change *or a 3-change* that produces a lower-cost tour.

---

[30] Whenever you've got a little extra time, it's worth passing the output of a heuristic algorithm through a local search postprocessing step. After all, the solution can only get better!

[31] In a 2-change, the pair of edges removed uniquely determines the pair to be added (Section 20.4.2). This is no longer the case for 3-changes. For example, if three edges with no shared endpoints are removed, there are seven ways to pair up their six endpoints that lead to new feasible tours (as you should check).

Quiz 20.10

Fix an instance of the TSP. Let H_2 and H_3 denote the meta-graphs corresponding to the 2OPT and 3OPT algorithms, respectively. Which of the following statements are true? (Choose all that apply.)

  a) Every edge of H_2 is also an edge of H_3.

  b) Every edge of H_3 is also an edge of H_2.

  c) Every local minimum of H_2 is also a local minimum of H_3.

  d) Every local minimum of H_3 is also a local minimum of H_2.

(See Section 20.5.8 for the solution and discussion.)

Choosing Your Neighborhood Size

When there are competing definitions of "local moves," which one should you use? This question is usually best answered empirically, by trying out several options. Quiz 20.10 does, however, illustrate a general advantage of large neighborhood sizes: More local moves means fewer lousy local optima for local search to get stuck at. The primary downside of larger neighborhood sizes is the slowdown in checking for improving local moves. For example, in the TSP, checking for an improving 2-change takes quadratic time (in the number of vertices) while checking for an improving 3-change takes cubic time. One approach to balancing these pros and cons is to use the largest neighborhood you can get away with subject to a target per-iteration running time (like one second or ten seconds).

20.5.4   Two Algorithm Design Decisions

Steps 4 and 5 of the local search paradigm supply the details missing from the generic local search algorithm in Section 20.5.2.

Step 4: Decide how to choose an initial feasible solution. Two simple ways to choose an initial solution are greedily and randomly. For example, in the TSP, the initial tour could be constructed by the nearest neighbor algorithm (Quiz 20.7), or by choosing a uniformly random order to visit the vertices. In the makespan minimization problem, the initial schedule could be constructed by the Graham or LPT algorithms, or by assigning each job independently to a uniformly random machine. For the maximum coverage problem, the initial solution could be the output of the GreedyCoverage algorithm or a uniformly random choice of k of the given subsets.

Why cast aside a perfectly good greedy heuristic algorithm in favor of a random solution? Because starting local search from a better initial solution does not necessarily lead to a better (or even equally good) local optimum. An ideal initialization procedure quickly produces a not-too-bad solution that leaves lots of opportunities for local improvement. Random initialization often fits the bill.

Step 5: Decide how to choose among multiple improving local moves. The generic local search algorithm in Section 20.5.2 does not specify how to choose one improving local move from many. The simplest approach is to enumerate local moves one by one until an improving one is found.[32] An alternative with a slower per-iteration running time but a larger per-iteration objective function improvement is to complete the enumeration and greedily execute the local move offering the biggest improvement. A third option is to encourage wider exploration of the solution space by choosing one of the improving moves at random.

20.5.5   Running Time and Solution Quality

Steps 1–5 fully specify a local search algorithm, which starts from an initial solution (chosen using the procedure from step 4) and repeatedly performs improving local moves (chosen using the procedure from step 5) until a local optimum is reached and no further improving moves are possible. What kind of performance can you expect from such an algorithm?

All the lessons learned about the 2OPT algorithm for the TSP in Sections 20.4.4–20.4.5 apply to most other local search algorithms:

---

[32]The 2OPT algorithm followed this approach in the example in Section 20.4.3 (assuming that it always scanned 2-changes from left to right in the figures).

Common Characteristics of Local Search

1. Guaranteed to halt. (Assuming that there is only a finite number of feasible solutions.)

2. Not guaranteed to halt in a polynomial (in the input size) number of iterations.

3. On realistic inputs, almost always halts within a tolerable number of iterations.

4. Can be interrupted at any time to return the last (and best) solution found.

5. Not guaranteed to return a local optimum with an objective function value that is close to the best possible.

6. On realistic inputs, often produces high-quality local optima, but sometimes produces low-quality local optima.

20.5.6   Avoiding Bad Local Optima

Low-quality local optima can impede the successful application of local search. How can you tweak a local search algorithm to better avoid them? While one fix is to increase the neighborhood size (see Section 20.5.3), perhaps the easiest workaround is to rely on randomization, choosing a random initial solution or random improving moves in each iteration. You can then run as many independent trials of your algorithm as you have time for, returning the best local optimum found by any of the trials.

A more drastic approach to avoiding bad local optima is to sometimes allow non-improving local moves. For example, in each iteration:

(i) From the current solution, choose a local move uniformly at random.

(ii) If the chosen local move is improving, perform it.

(iii) Otherwise, if the chosen local move makes the objective function worse by $\Delta \geq 0$, perform it with some probability $p(\Delta)$ that is decreasing in Δ, and otherwise do nothing. (One popular choice for the function $p(\Delta)$ is the exponential function $e^{-\lambda\Delta}$, where $\lambda > 0$ is a tunable parameter.)[33]

Local search algorithms that permit non-improving moves do not generally halt and should be interrupted after a target amount of computation time.

There's no end to the additional bells and whistles you can layer on top of the basic local search algorithm.[34] Two different genres of them are:

- *History-dependent neighborhoods.* Rather than fixing the allowable local moves once and for all, they could depend on the trajectory-so-far of the local search algorithm. For example, you might disallow local moves that seem to partially reverse the previous move, such as a 2-change using some of the same endpoints as the previous 2-change.[35] Rules of this type are particularly useful for avoiding cycles in local search algorithms that allow non-improving moves.

- *Maintaining a population of solutions.* The algorithm could maintain $k \geq 2$ feasible solutions at all times, rather than only one. Each iteration of the algorithm now generates k new feasible solutions from k old ones, for example, by keeping only the k best neighbors of the current k solutions, or by combining pairs of current solutions to create new ones.[36]

20.5.7   When Should You Use Local Search?

You know a lot of algorithm design paradigms; when is local search the first one to try? If your application checks several of the following boxes, local search is probably worth a shot.

---

[33]If you've heard of the "Metropolis algorithm" or "simulated annealing," both are based on this idea.

[34]For a deep dive, check out the book *Local Search in Combinatorial Optimization*, edited by Emile Aarts and Jan Karel Lenstra (Princeton University Press, 2003).

[35]If you've heard of "tabu search" or the "Lin-Kernighan variable-depth heuristic," both are related to this idea.

[36]If you've heard of "beam search" or "genetic algorithms," both are variants of this idea.

When to Use Local Search

1. You don't have enough time to compute an exact solution.

2. You're willing to give up on running time and approximate correctness guarantees.

3. You want an algorithm that is relatively easy to implement.

4. You already have a good heuristic algorithm but want to improve its output further in a postprocessing step.

5. You want an algorithm that can be interrupted at any time.

6. State-of-the-art mixed integer programming and satisfiability solvers (discussed in Sections 21.4–21.5) aren't good enough—either because your input sizes are too big for the solvers to handle, or because your problem can't be translated easily into the format they require.

And remember, to get the most out of local search, you have to experiment—with different neighborhoods, initialization strategies, local move selection strategies, extra bells and whistles, and so on.

20.5.8 Solutions to Quizzes 20.9–20.10

Solution to Quiz 20.9

Correct answer: (a). The meta-graph has one vertex per tour, for a total of $\frac{1}{2}(n-1)!$ (see Quiz 19.1). Each vertex of the meta-graph is adjacent to $n(n-3)/2$ other vertices (see footnote 25). The total number of edges is therefore

$$\frac{1}{2} \cdot \underbrace{\frac{(n-1)!}{2}}_{\#\text{ of vertices}} \cdot \underbrace{\frac{n(n-3)}{2}}_{\#\text{ of incident edges}} = \frac{n!(n-3)}{8};$$

the leading "$\frac{1}{2}$" term corrects for the double-counting of each meta-graph edge (once via each endpoint).

Solution to Quiz 20.10

Correct answers: (a),(d). The 3OPT algorithm can make whatever 2-change (or 3-change) it likes in each iteration. With every local move available to 2OPT available also to 3OPT, answer (a) is correct. Thus, (d) is correct as well: If a vertex has a neighbor in H_2 with better objective function value (showing that it is not a local minimum in H_2), this same neighbor shows that the vertex is not a local minimum in H_3, either.

In the example in Figure 20.3, the first tour in the third row cannot be improved by a 2-change but can be improved by a 3-change (as you should check). This shows that (b) and (c) are both incorrect.

The Upshot

☆ In the makespan minimization problem, the goal is to assign jobs to machines to minimize the makespan (the maximum machine load).

☆ Making a single pass over the jobs and scheduling each on the currently least loaded machine produces a schedule with makespan at most twice the minimum possible.

☆ Sorting the jobs first (from longest to shortest) improves the guarantee from 2 to 4/3.

☆ In the maximum coverage problem, the goal is to choose k of m subsets to maximize their coverage (the size of their union).

☆ Greedily selecting subsets that increase the coverage as much as possible achieves a coverage of at least 63.2% of the maximum possible.

☆ The influence of a set of initially active vertices in a directed graph is the expected number of eventually activated vertices, assuming that an activated vertex activates each of its out-neighbors with some probability p.

☆ In the influence maximization problem, the goal is to choose k vertices of a directed graph to maximize their influence.

☆ Because influence is a weighted average of coverage functions, the 63.2% guarantee carries over to the greedy algorithm that iteratively selects vertices that increase the influence the most.

☆ In the traveling salesman problem (TSP), the input is a complete graph with real-valued edge costs, and the goal is to compute a tour (a cycle visiting every vertex exactly once) with the minimum-possible sum of edge costs.

☆ A 2-change creates a new tour from an old one by swapping out one pair of edges for another.

☆ The 2-OPT algorithm for the TSP repeatedly improves an initial tour via 2-changes until no such improvements are possible.

☆ A local search algorithm takes a walk through a meta-graph in which vertices correspond to feasible solutions (labeled by objective function value) and edges to local moves.

☆ A local search algorithm is specified by a meta-graph, an initial solution, and a rule for selecting among improving local moves.

☆ Local search algorithms often produce high-quality solutions in a reasonable amount of time, despite lacking provable running time and approximate correctness guarantees.

☆ Local search algorithms can be tweaked to better avoid low-quality local optima, for example, by allowing randomization and non-improving local moves.

Test Your Understanding

Problem 20.1 *(S)* In the makespan minimization problem (Section 20.1.1), suppose that jobs have similar lengths (with $\ell_j \leq 2\ell_h$ for all jobs j, h) and that there is a healthy number of jobs (at least 10 times the number of machines). What can you say about the makespan of the schedule output by the Graham algorithm of Section 20.1.3? (Choose the strongest true statement.)

a) It is at most 10% larger than the minimum-possible makespan.

b) It is at most 20% larger than the minimum-possible makespan.

c) It is at most 50% larger than the minimum-possible makespan.

d) It is at most 100% larger than the minimum-possible makespan.

Problem 20.2 *(S)* The goal in the maximum coverage problem (Section 20.2.1) is to cover as many elements as possible using a fixed number of subsets; in the closely related *set cover* problem, the goal is to cover *all* the elements while using as few subsets as possible (like hiring a team with all the requisite skills at the minimum-possible cost).[37] The greedy algorithm for the maximum coverage problem (Section 20.2.3) extends easily to the set cover problem (given as input m subsets $A_1, A_2, \ldots, A_m$ of a ground set U, with $\cup_{i=1}^m A_i = U$):

Greedy Heuristic Algorithm for Set Cover

$K := \emptyset$ // indices of chosen sets
while $f_{cov}(K) < |U|$ **do**   // part of U uncovered
     $i^* := \text{argmax}_{i=1}^m \left[f_{cov}(K \cup \{i\}) - f_{cov}(K) \right]$
     $K := K \cup \{i^*\}$
return K

Let k denote the minimum number of subsets required to cover all of U. Which of the following approximate correctness guarantees holds for this algorithm? (Choose the strongest true statement.)

a) Its solution consists of at most $2k$ subsets.

---
[37] This problem is NP-hard; see Problem 22.6.

b) Its solution consists of $O(k \log |U|)$ subsets.

c) Its solution consists of $O(k \cdot \sqrt{|U|})$ subsets.

d) Its solution consists of $O(k \cdot |U|)$ subsets.

Problem 20.3 *(S)* This problem considers three greedy heuristics for the knapsack problem (defined in Section 19.4.2). The input consists of n items with values $v_1, v_2, \ldots, v_n$ and sizes $s_1, s_2, \ldots, s_n$, and a knapsack capacity C.

Greedy Heuristic Algorithm #1 for Knapsack

$I := \emptyset, \, S := 0$ // chosen items and their size
sort and reindex the jobs so that $v_1 \geq v_2 \geq \cdots \geq v_n$
for $i = 1$ to n **do**
     if $S + s_i \leq C$ **then** // choose item if feasible
         $I := I \cup \{i\}, \, S := S + s_i$
return I

Greedy Heuristic Algorithm #2 for Knapsack

$I := \emptyset, \, S := 0$ // chosen items and their size
sort and reindex the jobs so that $\frac{v_1}{s_1} \geq \frac{v_2}{s_2} \geq \cdots \geq \frac{v_n}{s_n}$
for $i = 1$ to n **do**
     if $S + s_i \leq C$ **then** // choose item if feasible
         $I := I \cup \{i\}, \, S := S + s_i$
return I

Greedy Heuristic Algorithm #3 for Knapsack

$I_1 :=$ output of greedy heuristic algorithm #1
$I_2 :=$ output of greedy heuristic algorithm #2
return whichever of I_1, I_2 has higher total value

Which of the following statements are true? (Choose all that apply.)

a) The total value of the solution returned by the first greedy algorithm is always at least 50% of the maximum possible.

b) The total value of the solution returned by the second greedy algorithm is always at least 50% of the maximum possible.

c) The total value of the solution returned by the third greedy algorithm is always at least 50% of the maximum possible.

d) If every item size is at most 10% of the knapsack capacity (that is, $\max_{i=1}^{n} s_i \leq C/10$), the total value of the solution returned by the first greedy algorithm is at least 90% of the maximum possible.

e) If every item size is at most 10% of the knapsack capacity, the total value of the solution returned by the second greedy algorithm is at least 90% of the maximum possible.

f) If every item size is at most 10% of the knapsack capacity, the total value of the solution returned by the third greedy algorithm is at least 90% of the maximum possible.

Problem 20.4 *(S)* In the *vertex cover* problem, the input is an undirected graph $G = (V, E)$, and the goal is to identify a minimum-size subset $S \subseteq V$ of vertices that includes at least one endpoint of every edge in E.[38] (For example, perhaps the edges represent roads and the vertices intersections, and the goal is to monitor all the roads while installing security cameras at as few intersections as possible.) One simple heuristic algorithm repeatedly chooses a not-yet-covered edge and adds *both* its endpoints to its solution-so-far:

Heuristic Algorithm for Vertex Cover

$S := \emptyset$ // chosen vertices
while there is an edge $(v, w) \in E$ with $v, w \notin S$ **do**
     $S := S \cup \{v, w\}$ // add both endpoints of edge
return S

---

[38] This problem is NP-hard; see Problem 22.5.

Let k denote the minimum number of vertices required to capture at least one endpoint of each edge. Which of the following approximate correctness guarantees holds for this algorithm? (Choose the strongest true statement.)

a) Its solution consists of at most $2k$ vertices.

b) Its solution consists of $O(k \log |E|)$ vertices.

c) Its solution consists of $O(k \cdot \sqrt{|E|})$ vertices.

d) Its solution consists of $O(k \cdot |E|)$ vertices.

Problem 20.5 *(S)* Which of the following statements about the generic local search algorithm in Section 20.5.2 is *not* true?

a) Its output generally depends on the choice of the initial feasible solution.

b) Its output generally depends on the method for choosing one improving local move from many.

c) It will always, eventually, halt at an optimal solution.

d) In some cases, it performs an exponential (in the input size) number of iterations before halting.

Challenge Problems

Problem 20.6 *(H)* Propose an implementation of the `Graham` algorithm (Section 20.1.3) that uses a heap data structure and runs in $O(n \log m)$ time, where n is the number of jobs and m is the number of machines.[39]

Problem 20.7 *(H)* This problem improves Theorem 20.4 and extends the example in Quiz 20.3 to identify the best-possible approximate correctness guarantee for the LPT algorithm (Section 20.1.7).

---

[39]Technically, the running time will be $O(m + n \log m)$. The problem is uninteresting when $n \leq m$, however, as in that case each job can be granted a dedicated machine.

(a) Let job j be the last job assigned to the most heavily loaded machine in the schedule returned by the LPT algorithm. Prove that if $\ell_j > M^*/3$, where M^* denotes the minimum-possible makespan, then this schedule is optimal (that is, has makespan M^*).

(b) Prove that the LPT algorithm always outputs a schedule with makespan at most $\frac{4}{3} - \frac{1}{3m}$ times the minimum possible, where m denotes the number of machines.

(c) Generalize the example in Quiz 20.3 to show that, for every $m \geq 1$, there is an example with m machines in which the schedule produced by the LPT algorithm has makespan $\frac{4}{3} - \frac{1}{3m}$ times the minimum possible.

Problem 20.8 *(H)* Recall the bad example for the GreedyCoverage algorithm in Quiz 20.5.

(a) Prove Proposition 20.6.

(b) Extend your examples in (a) to show that, even with best-case tie-breaking, for every constant $\epsilon > 0$, the GreedyCoverage algorithm does not guarantee a $1 - (1 - \frac{1}{k})^k + \epsilon$ fraction of the maximum-possible coverage (where k denotes the number of subsets chosen).

Problem 20.9 *(H)* Show that every instance of the maximum coverage problem can be encoded as an instance of the influence maximization problem so that: (i) the two instances have the same optimal objective function value F^*; and (ii) any solution to the latter instance with influence F can be easily converted to a solution to the former instance with coverage at least F.

Problem 20.10 *(H)* The goal in the maximum coverage problem is to choose k subsets to maximize the coverage f_{cov}. The goal in the influence maximization problem is to choose k vertices to maximize the influence f_{inf}. The general version of this type of problem is: Given a set O of objects and a real-valued set function f (specifying a number $f(S)$ for each subset $S \subseteq O$), choose k objects of O to maximize f. The GreedyCoverage and GreedyInfluence algorithms extend naturally to the general problem:

Greedy Algorithm for Set Function Maximization

$S := \emptyset$ `// chosen objects`
for $j = 1$ to k **do**   `// choose objects one by one`
    `// greedily increase objective function`
    $o^* := \text{argmax}_{o \notin S} \left[f(S \cup \{o\}) - f(S) \right]$
    $S := S \cup \{o^*\}$
return S

For which objective functions f does this greedy algorithm enjoy an approximate correctness guarantee akin to Theorems 20.7 and 20.9? Here are the key properties:

1. *Nonnegative:* $f(S) \geq 0$ for all $S \subseteq O$.

2. *Monotone:* $f(S) \geq f(T)$ whenever $S \supseteq T$.

3. *Submodular:* $f(S \cup \{o\}) - f(S) \leq f(T \cup \{o\}) - f(T)$ whenever $S \supseteq T$ and $o \notin S$.[40]

(a) Prove that the coverage and influence functions f_{cov} and f_{inf} possess all three properties.

(b) Prove that whenever f is nonnegative, monotone, and submodular, the general greedy heuristic algorithm is guaranteed to return a set S of objects that satisfies

$$f(S) \geq \left(1 - \left(1 - \frac{1}{k} \right)^k \right) \cdot f(S^*),$$

where S^* maximizes f over all size-k subsets of O.

Problem 20.11 *(H)* Problem 20.3 investigated approximate correctness guarantees for greedy heuristic algorithms for the knapsack problem. This problem outlines a dynamic programming algorithm with a much stronger guarantee: For a user-specified error parameter $\epsilon > 0$ (like .1 or .01), the algorithm outputs a solution with total value at

---

[40]Submodularity asserts a "diminishing returns" property: The marginal value of a new object o can only diminish as other objects are acquired.

least $1 - \epsilon$ times the maximum possible. (Full disclosure, in case this sounds too good to be true for an NP-hard problem: The running time of the algorithm blows up as ϵ approaches 0.)

(a) Section 19.4.2 mentioned that the knapsack problem can be solved in $O(nC)$ time using dynamic programming, where n denotes the number of items and C the knapsack capacity; see also Chapter 16 of *Part 3*. (All item values and sizes, as well as the knapsack capacity, are positive integers.) Give a different dynamic programming algorithm for the problem that runs in $O(n^2 \cdot v_{max})$ time, where v_{max} denotes the largest value of any item.

(b) To shrink the item values down to a manageable magnitude, divide each of them by $m := (\epsilon \cdot v_{max})/n$ and round each result down to the nearest integer (where ϵ is the user-specified error parameter). Prove that the total value of every feasible solution goes down by at least a factor of m and that the total value of an optimal solution goes down by at most a factor of $m/(1 - \epsilon)$. (You can assume that every item has size at most C and hence fits in the knapsack.)

(c) Propose an $O(n^3/\epsilon)$-time algorithm that is guaranteed to return a feasible solution with total value at least $1 - \epsilon$ times the maximum possible.[41]

Problem 20.12 *(H)* This problem describes a commonly encountered special case of the traveling salesman problem for which there are fast heuristic algorithms with good approximate correctness guarantees. In a *metric* instance $G = (V, E)$ of the TSP, all the edge costs c_e are nonnegative and the shortest path between any two vertices is the direct one-hop path (a condition that is known as the "triangle inequality"):

$$c_{vw} \leq \sum_{e \in P} c_e$$

for every pair $v, w \in V$ of vertices and v-w path P. (The example in Quiz 19.2 is a metric instance, while the example in Quiz 20.7 is not.)

---

[41] A heuristic algorithm with this type of guarantee is called a *fully polynomial-time approximation scheme (FPTAS)*.

The triangle inequality typically holds in applications in which edge costs correspond to physical distances. The TSP remains NP-hard in the special case of metric instances (see Problem 22.12(a)).

Our starting point for a fast heuristic algorithm is the linear-time algorithm for tree instances described in Problem 19.8. The key idea is to reduce a general metric instance to a tree instance by computing a minimum spanning tree.

MST Heuristic for Metric TSP

$T :=$ minimum spanning tree of the input graph G
return an optimal tour of the tree instance defined by T

The first step can be implemented in near-linear time using Prim's or Kruskal's algorithm. The second step can be implemented in linear time using the solution to Problem 19.8. In the tree TSP instance constructed in the second step, the length a_e of an edge e of T is set to the cost c_e of that edge in the given metric TSP instance (with the cost of each edge (v, w) in the tree instance then defined as the total length $\sum_{e \in P_{vw}} a_e$ of the unique v-w path P_{vw} in T):

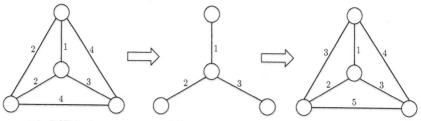

metric TSP instance   a minimum spanning tree   tree TSP instance

(a) Prove that the minimum total cost of a traveling salesman tour is at least that of a minimum spanning tree. (This step does not require the triangle inequality.)

(b) Prove that, for every instance of metric TSP, the total cost of the tour computed by the MST heuristic is at most twice the minimum possible.

Problem 20.13 *(H)* Propose an implementation of the 2OPT algorithm (Section 20.4.3) in which each iteration of the main while loop runs in $O(n^2)$ time, where n is the number of vertices.

Problem 20.14 *(S)* Most local search algorithms lack polynomial running time and approximate correctness guarantees; this problem describes a rare exception. For an integer $k \geq 2$, in the *maximum k-cut* problem, the input is an undirected graph $G = (V, E)$. The feasible solutions are the k-cuts of the graph, meaning partitions of the vertex set V into k non-empty groups $S_1, S_2, \ldots, S_k$. The objective is to maximize the number of edges with endpoints in different groups. For example, in the graph

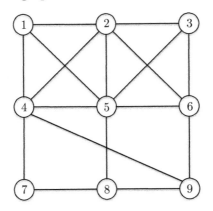

sixteen of the seventeen edges have endpoints in distinct groups of the 3-cut $(\{1, 6, 7\}, \{2, 5, 9\}, \{3, 4, 8\})$.[42]

For a k-cut $(S_1, S_2, \ldots, S_k)$, each local move corresponds to a reassignment of a single vertex from one group to another, subject to the constraint that none of the k groups can become empty.

(a) Prove that, for every initial k-cut and selection rule for choosing improving local moves, the generic local search algorithm halts within $|E|$ iterations.

(b) Prove that, for every initial k-cut and selection rule for choosing improving local moves, the generic local search algorithm halts with a k-cut with an objective function value of at least $(k-1)/k$ times the maximum possible.

---

[42]There's no way to do better, as two of the vertices in $\{1, 2, 4, 5\}$ must belong to a common group.

Programming Problems

Problem 20.15 Implement in your favorite programming language the nearest neighbor algorithm for the TSP (as seen in Quiz 20.7). Try out your implementation on instances with edge costs chosen independently and uniformly at random from the set $\{1, 2, \ldots, 100\}$ or, alternatively, for vertices that correspond to points chosen independently and uniformly at random from the unit square.[43] How large an input size (that is, how many vertices) can your program reliably process in under a minute? What about in under an hour? (See www.algorithmsilluminated.org for test cases and challenge data sets.)

Problem 20.16 Implement in your favorite programming language the 2OPT algorithm from Section 20.4.3. Use your implementation of the nearest neighbor algorithm from Problem 20.15 to compute the initial tour. Implement each iteration of the main loop so that it runs in quadratic time (see Problem 20.13) and experiment with different ways of selecting an improving local move. Try out your implementation on the same instances you used for Problem 20.15.[44] By how much does local search improve the total cost of the initial tour? Which of your selection rules leads to the most dramatic improvement? (See www.algorithmsilluminated.org for test cases and challenge data sets.)

---

[43]That is, the x- and y-coordinates of each point are independent and uniformly random numbers in $[0, 1]$. The cost of the edge connecting two points (x_1, y_1) and (x_2, y_2) is then defined as the Euclidean (that is, straight-line) distance between them, which is $\sqrt{(x_1 - x_2)^2 + (y_1 - y_2)^2}$. (For the nearest neighbor algorithm, you can equivalently work with squared Euclidean distances.)

[44]For points in the unit square, does it matter whether you use Euclidean or squared Euclidean distances?

Chapter 21

Compromising on Speed:
Exact Inefficient Algorithms

You can't have it all with NP-hard problems. When correctness cannot be compromised and heuristic algorithms are out of the question, it's time to consider correct algorithms that do not always run in polynomial time. The goal is then to design a general-purpose and correct algorithm that is as fast as possible, and certainly faster than exhaustive search, on as many inputs as possible. Sections 21.1 and 21.2 use dynamic programming to design algorithms that are always faster than exhaustive search in two case studies: the TSP and the problem of finding a long path in a graph. Sections 21.3–21.5 introduce mixed integer programming and satisfiability solvers, which lack better-than-exhaustive-search running time guarantees but can nevertheless be highly effective at solving the instances of NP-hard problems that arise in practice.

21.1 The Bellman-Held-Karp Algorithm for the TSP

21.1.1 The Baseline: Exhaustive Search

In the TSP (Section 19.1.2), the input is a complete graph $G = (V, E)$ with real-valued edge costs and the goal is to compute a tour—a cycle visiting every vertex exactly once—with the minimum-possible sum of edge costs. The TSP is NP-hard (see Section 22.7); if correctness cannot be compromised, the only option is to resort to algorithms that run in super-polynomial (and presumably exponential) time in the worst case (assuming, as usual, that the P $\neq$ NP conjecture is true). Can algorithmic ingenuity at least improve over mindless exhaustive search? How big a speed-up can we hope for?

Solving the TSP by exhaustive search through the $\frac{1}{2}(n - 1)!$ possible tours (Quiz 19.1) results in an $O(n!)$-time algorithm. The factorial function $n! = n \cdot (n-1) \cdot (n-2) \cdots 2 \cdot 1$ certainly grows more quickly than a simple exponential function like 2^n—the latter is the product of n 2's, the former the product of n terms that are mostly much bigger than 2. Just how big is it? There's a remarkably accurate answer to this question, called *Stirling's approximation*.[1] (In it, $e = 2.718\ldots$ denotes Euler's number and, of course, $\pi = 3.14\ldots$.)

Stirling's Approximation

$$n! \approx \sqrt{2\pi n}\left(\frac{n}{e}\right)^n \tag{21.1}$$

Stirling's approximation shows that the factorial function (with its n^n-type dependence) grows a *lot* faster than 2^n. For example, you could run to completion an $n!$-time algorithm on modern computers only for n up to maybe 15, while a 2^n-time algorithm could handle input sizes up to $n = 40$ or so. (Still not that impressive, perhaps, but you take what you can get with NP-hard problems!) Thus, solving the TSP in time closer to 2^n than $n!$ is a worthy goal.

21.1.2 Dynamic Programming

While many of the killer applications of dynamic programming are to polynomial-time solvable problems, the paradigm is equally adept at solving NP-hard problems faster than exhaustive search, including the knapsack problem (Section 19.4.2), the TSP (this section), and more (Section 21.2). To review (for example, from Chapter 16 of *Part 3*), the dynamic programming paradigm is:

The Dynamic Programming Paradigm

1. Identify a relatively small collection of subproblems.

2. Show how to quickly and correctly solve "larger" subproblems given the solutions to "smaller" ones.

3. Show how to quickly and correctly infer the final solu-

---

[1]Remember the name but not the formula or its proof; you can always look it up on Wikipedia or elsewhere when you need it.

> tion from the solutions to all the subproblems.

After these three steps are implemented, the corresponding dynamic programming algorithm writes itself: Systematically solve all the subproblems one by one, working from "smallest" to "largest," and extract the final solution from those of the subproblems.

For example, suppose a dynamic programming algorithm solves at most $f(n)$ different subproblems (working systematically from "smallest" to "largest"), using at most $g(n)$ time for each, and performs at most $h(n)$ postprocessing work to extract the final solution (where n denotes the input size). The algorithm's running time is then at most

$$\underbrace{f(n)}_{\substack{\text{\# subproblems}}} \quad \times \quad \underbrace{g(n)}_{\substack{\text{time per subproblem} \\ \text{(given previous solutions)}}} \quad + \quad \underbrace{h(n)}_{\substack{\text{postprocessing}}} \ . \qquad (21.2)$$

When applying dynamic programming to an NP-hard problem like the TSP, we should expect at least one of the functions $f(n)$, $g(n)$, or $h(n)$ to be exponential in n. Looking back at some canonical dynamic programming algorithms, we can see that the functions $g(n)$ and $h(n)$ are almost always $O(1)$ or $O(n)$, while the number $f(n)$ of subproblems varies widely from algorithm to algorithm.[2] We should, therefore, be ready for a dynamic programming algorithm for the TSP to use an exponential number of subproblems.

21.1.3   Optimal Substructure

The key that unleashes the potential of dynamic programming is the identification of the right collection of subproblems. The best way to home in on them is to think through the different ways that an optimal solution might be built up from optimal solutions to smaller subproblems.

Suppose someone handed us on a silver platter a minimum-cost traveling salesman tour T of the vertices $V = \{1, 2, \ldots, n\}$, with

---

[2]For example, the dynamic programming algorithm for the weighted independent set problem in path graphs solves $O(n)$ subproblems (where n denotes the number of vertices), while that for the knapsack problem solves $O(nC)$ subproblems (where n denotes the number of items and C the knapsack capacity). The Bellman-Ford and Floyd-Warshall shortest-path algorithms use $O(n^2)$ and $O(n^3)$ subproblems, respectively (where n denotes the number of vertices).

$n \geq 3$. What must it look like? In how many different ways could it have been built up from optimal solutions to smaller subproblems? Think of the tour T as starting and ending at the vertex 1, and zoom in on its last decision—its final edge, from some vertex j back to the starting point 1. If we only knew the identity of j, we would know what the tour looks like: a minimum-cost cycle-free path from 1 to j that visits every vertex, followed by the edge from j back to 1:[3]

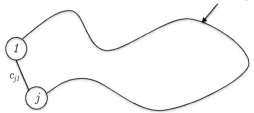

1-j path (visits every vertex, cycle-free, min-cost)

Thus, there are $n-1$ and only $n-1$ candidates vying to be an optimal traveling salesman tour (one for each choice $j \in \{2, 3, \ldots, n\}$ of the final vertex), and the best of these must be a minimum-cost tour:[4]

$$\text{optimal tour cost} = \min_{j=2}^{n} \left(\begin{array}{c} \text{min cost of cycle-} \\ \text{free 1-}j \text{ path that} \\ \text{visits every vertex} \end{array} + c_{j1} \right). \quad (21.3)$$

So far, so good. But the argument grows trickier if we forge ahead. Consider an optimal solution to one of our $n-1$ subproblems—a minimum-cost path from 1 to j that visits every vertex exactly once (equivalently, that visits every vertex and is cycle-free). What must it look like?

Quiz 21.1

Let P be a minimum-cost cycle-free path from 1 to j that visits every vertex, with final hop (k, j). Let P' denote P

---

[3]Why must $T - \{(j, 1)\}$ be such a minimum-cost path? Because if there were a lower-cost cycle-free 1-j path visiting every vertex, we could plug the edge $(j, 1)$ back in to recover a lower-cost tour (contradicting the optimality of T).

[4]Thinking recursively, a minimum-cost tour can be computed by iterating over the $n-1$ choices for the vertex j and, in each iteration, recursively computing a minimum-cost cycle-free 1-j path that visits every vertex.

with its final hop (k, j) removed. Which of the following are true? (Choose all that apply.)

   a) P' is a cycle-free path from 1 to k that visits every vertex of $V - \{j\}$.

   b) P' is a minimum-cost path of the form in (a).

   c) P' is a cycle-free path from 1 to k that visits every vertex of $V - \{j\}$ and does not visit vertex j.

   d) P' is a minimum-cost path of the form in (c).

(See Section 21.1.7 for the solution and discussion.)

The solution to Quiz 21.1 proves the following lemma.

Lemma 21.1 (TSP Optimal Substructure) *Assume that $n \geq 3$. Suppose P is a minimum-cost cycle-free path from vertex 1 to vertex j that visits every vertex of $V = \{1, 2, \ldots, n\}$, and its final hop is (k, j). The 1-k subpath P' is a minimum-cost cycle-free 1-k path that visits exactly the vertices $V - \{j\}$.*

In other words, once you know the last hop of an optimal path, you know what the rest of it must look like.

The subproblem solved optimally by P' in Lemma 21.1 specifies the exact subset of vertices to visit. The bad news is that this will force our dynamic programming algorithm to use subproblems indexed by subsets of vertices (of which there are, unfortunately, an exponential number). The good news is that these subproblems will not specify the *order* in which to visit the vertices. For this reason, their number will scale with 2^n rather than $n!$.[5]

21.1.4   Recurrence

Lemma 21.1 narrows down the possibilities for an optimal path from vertex 1 to a vertex j to $n - 2$ and only $n - 2$ candidates—one for each choice of the penultimate vertex k. The best of these $n - 2$ candidates must be an optimal path.

---

[5]For the same reason, the *memory* required by the algorithm will also scale with 2^n (unlike exhaustive search, which uses minimal memory).

Corollary 21.2 (TSP Recurrence) *With the assumptions and notation of Lemma 21.1, let $C_{S,j}$ denote the minimum cost of a cycle-free path that begins at the vertex 1, ends at the vertex $j \in S$, and visits exactly the vertices in the subset $S \subseteq V$. Then, for every $j \in V - \{1\}$,*

$$C_{V,j} = \min_{\substack{k \in V \\ k \neq 1, j}} \left(C_{V-\{j\},k} + c_{kj}\right). \tag{21.4}$$

More generally, for every subset $S \subseteq V$ that contains 1 and at least two other vertices, and for every vertex $j \in S - \{1\}$,

$$C_{S,j} = \min_{\substack{k \in S \\ k \neq 1, j}} \left(C_{S-\{j\},k} + c_{kj}\right). \tag{21.5}$$

The second statement in Corollary 21.2 follows by applying the first statement to the vertices of S, viewed as a TSP instance in their own right (with edge costs inherited from the original instance). The "min" in the recurrences (21.4) and (21.5) implements exhaustive search over the candidates for the penultimate vertex of an optimal solution.

21.1.5 The Subproblems

Ranging over all relevant values of the parameters S and j in the recurrence (21.5), we obtain our collection of subproblems. The base cases correspond to subsets of the form $\{1, j\}$ for some $j \in V - \{1\}$.[6]

TSP: Subproblems

Compute $C_{S,j}$, the minimum cost of a cycle-free path from vertex 1 to vertex j that visits exactly the vertices in S.

(For each $S \subseteq \{1, 2, \ldots, n\}$ containing vertex 1 and at least one other vertex, and each $j \in S - \{1\}$.)

The identity in (21.3) shows how to compute the minimum cost of a tour from the solutions to the largest subproblems (with $S = V$):

$$\text{optimal tour cost} = \min_{j=2}^{n} \left(C_{V,j} + c_{j1}\right). \tag{21.6}$$

---

[6]Thinking recursively, each application of the recurrence (21.5) effectively removes one vertex (other than vertex 1) from further consideration. These vertex choices are arbitrary, so we must be prepared for any subset of vertices (that contains 1 and at least one other vertex).

21.1.6   The Bellman-Held-Karp Algorithm

With the subproblems, recurrence (21.5), and postprocessing step (21.6) in hand, the dynamic programming algorithm for the TSP writes itself. There are $2^{n-1} - 1$ choices of S to keep track of (one per non-empty subset of $\{2, 3, \ldots, n\}$), and "subproblem size" is measured by the number of vertices to visit (the size of S). For a base case with subset $S = \{1, j\}$, the only option is the one-hop 1-j path with cost c_{1j}. In the following pseudocode, the subproblem array is indexed by vertex subsets S; a concrete implementation would encode these subsets by integers.[7]

BellmanHeldKarp

Input: complete undirected graph $G = (V, E)$ with $V = \{1, 2, \ldots, n\}$ and $n \geq 3$, and a real-valued cost c_{ij} for each edge $(i, j) \in E$.

Output: the minimum total cost of a traveling salesman tour of G.

```
// subproblems (1 ∈ S, |S| ≥ 2, j ∈ V − {1})
// (only subproblems with j ∈ S are ever used)
```
$A := (2^{n-1} - 1) \times (n - 1)$ two-dimensional array

```
// base cases (|S| = 2)
```
for $j = 2$ **to** n **do**
    $A[\{1, j\}][j] := c_{1j}$

```
// systematically solve all subproblems
```
for $s = 3$ **to** n **do**   `// s=subproblem size`
    for S with $|S| = s$ and $1 \in S$ **do**
        for $j \in S - \{1\}$ **do**
            `// use recurrence from Corollary 21.2`
            $A[S][j] := \min_{\substack{k \in S \\ k \neq 1, j}} (A[S - \{j\}][k] + c_{kj})$

```
// use (21.6) to compute the optimal tour cost
```
return $\min_{j=2}^{n} (A[V][j] + c_{j1})$

---

[7]For example, the subsets of $V - \{1\}$ can be represented using length-$(n - 1)$ bit arrays, which in turn can be interpreted as the binary expansions of the integers between 0 and $2^{n-1} - 1$.

In the loop iteration responsible for computing the subproblem solution $A[S][j]$, all terms of the form $A[S - \{j\}][k]$ have been computed in the previous iteration of the outermost for loop (or in the base cases). These values are ready and waiting to be looked up in constant time.[8,9]

The correctness of the BellmanHeldKarp algorithm follows by induction (on the subproblem size), with the recurrence in Corollary 21.2 justifying the inductive step and the identity (21.6) the final postprocessing step.[10]

And the running time? The base cases and the postprocessing step take $O(n)$ time. There are $(2^{n-1} - 1)(n - 1) = O(n2^n)$ subproblems. Solving a subproblem boils down to the minimum computation in the inner loop, which takes $O(n)$ time. The overall running time is then $O(n^2 2^n)$.[11,12]

Theorem 21.3 (Properties of BellmanHeldKarp) *For every complete graph $G = (V, E)$ with $n \geq 3$ vertices and real-valued edge costs, the BellmanHeldKarp algorithm runs in $O(n^2 2^n)$ time and returns the minimum cost of a traveling salesman tour.*

The BellmanHeldKarp algorithm computes the total cost of an optimal tour, not an optimal tour itself. As always with dynamic programming algorithms, you can reconstruct an optimal solution in a postprocessing step that traces back through the filled-in subproblem array (Problem 21.6).

21.1.7   Solution to Quiz 21.1

Correct answers: (a),(c),(d). Because P is a cycle-free path from 1 to j visiting every vertex and with final hop (k, j), the subpath P'

---

[8]This algorithm was proposed independently by Richard E. Bellman in the paper "Dynamic Programming Treatment of the Travelling Salesman Problem" (*Journal of the ACM*, 1962) and Michael Held and Richard M. Karp in the paper "A Dynamic Programming Approach to Sequencing Problems" (*Journal of the Society for Industrial and Applied Mathematics*, 1962).

[9]For an example of the BellmanHeldKarp algorithm in action, see Problem 21.2.

[10]For an induction refresher, see Appendix A of *Part 1* or the resources at www.algorithmsilluminated.org.

[11]In the notation of (21.2), $f(n) = O(n2^n)$, $g(n) = O(n)$, and $h(n) = O(n)$.

[12]This running time analysis assumes that the subsets S with a given size $s \geq 3$ and $1 \in S$ are enumerated in time linear in their number $\binom{n-1}{s-1}$, for example, by recursive enumeration. (If you want to venture out into the weeds on this point, look up "Gosper's hack.")

is a cycle-free path from 1 to k that visits all the vertices of $V - \{j\}$ but not j. Thus, (a) and (c) are both correct answers. Answer (b) is incorrect because P' might not be able to compete with the cycle-free paths from 1 to k that visit every vertex of $V - \{j\}$ and *can also* visit j:

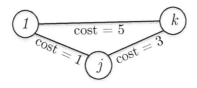

We can prove (d) by contradiction.[13] Let C denote the total cost of P, so that the cost of P' is $C - c_{kj}$. If (d) is false, there is another cycle-free path P^* from 1 to k that visits every vertex of $V - \{j\}$, does not visit vertex j, and has cost $C^* < C - c_{kj}$. Then, appending the edge (k, j) to P^* produces a path $\widehat{P}$ from 1 to j with total cost $C^* + c_{kj} < C$:

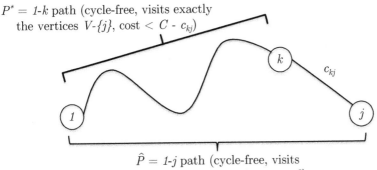

$$P^* = \textit{1-k path (cycle-free, visits exactly the vertices } V\text{-}\{j\}, \text{ cost} < C - c_{kj})$$

$$\widehat{P} = \textit{1-j path (cycle-free, visits exactly the vertices } V, \text{ cost} < C)$$

Moreover, the path $\widehat{P}$ is cycle-free (because P^* is cycle-free and does not visit j) and visits every vertex of V (because P^* visits every vertex of $V - \{j\}$). This contradicts our assumption that P is a minimum-cost such path.

---

[13] Recall that in this type of proof, you assume the *opposite* of what you want to prove, and then build on this assumption with a sequence of logically correct steps that culminates in a patently false statement. Such a contradiction implies that the assumption can't be true, which proves the desired statement.

*21.2   Finding Long Paths by Color Coding

Graphs are omnipresent in the study of algorithms because they hit the sweet spot between expressiveness and tractability. Throughout this book series, we've seen many efficient algorithms for processing graphs (graph search, connected components, shortest paths, etc.), and many application domains well modeled by graphs (road networks, the World Wide Web, social networks, etc.). This section furnishes another example, a killer application of dynamic programming and randomization to the detection of structure in biological networks.

21.2.1   Motivation

Most of the work that takes place in a cell is carried out by proteins (chains of amino acids), often acting in concert. For example, a series of proteins might transmit a signal that arrives at the cell membrane to the proteins that regulate the transcription of the cell's DNA to RNA. Understanding such signaling pathways and how they get rewired by genetic mutations is an important step in developing new drugs to combat diseases.

Interactions between proteins are naturally modeled as a graph, called a *protein-protein interaction (PPI)* network, with one vertex per protein and one edge per pair of proteins suspected to interact. The simplest signaling pathways are *linear* pathways, corresponding to paths in the PPI network. How quickly can we find them?

21.2.2   Problem Definition

The problem of finding a linear pathway of a given length in a PPI network can be cast as the following minimum-cost k-path problem, where a *k-path* of a graph is a (cycle-free) path of $k-1$ edges visiting k distinct vertices.

Problem: Minimum-Cost k-Path

Input: An undirected graph $G = (V, E)$, a real-valued cost c_e for each edge $e \in E$, and a positive integer k.

Output: A k-path P of G with the minimum-possible total cost $\sum_{e \in P} c_e$. (Or if G has no k-paths, report this fact.)

The edge costs reflect the uncertainties inevitable in noisy biological data, with a higher cost indicating a lower confidence that the corresponding pair of proteins really do interact. (Missing edges effectively have a cost of $+\infty$.) In a PPI network, the minimum-cost k-path corresponds to the most plausible linear pathway of a given length. In realistic instances, k might be 10 or 20; the number of vertices might be in the hundreds or thousands.

For example, in the graph

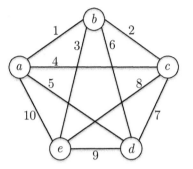

the minimum cost of a 4-path is 8 ($c \to a \to b \to e$).

The minimum-cost k-path problem is closely related to the TSP and, for this reason, is NP-hard (see Section 22.3). But can we at least improve over exhaustive search?

21.2.3   A First Stab at the Subproblems

The minimum-cost k-path problem closely resembles the TSP, with the main difference being the path length bound k. Why not use the same subproblems that served us so well in beating exhaustive search for the TSP (Section 21.1.5)? That is, given a graph $G = (V, E)$ with real-valued edge costs and a path length k:

Subproblems (A First Stab)

Compute $C_{S,v}$, the minimum cost of a cycle-free path that ends at the vertex $v \in V$ and visits exactly the vertices in S (or $+\infty$, if there is no such path).

(For each non-empty subset $S \subseteq V$ of at most k vertices and each $v \in S$.)

Because a minimum-cost k-path could start anywhere, subproblems no longer specify a starting vertex (which in the TSP was always vertex 1). The minimum cost of a k-path is the smallest of the solutions to the biggest subproblems (with $|S| = k$); if the graph has no k-paths, all such subproblem solutions will be $+\infty$.

Quiz 21.2

Suppose $k = 10$. How many subproblems are there, as a function of the number of vertices n? (Choose the strongest correct statement.)

a) $O(n)$

b) $O(n^{10})$

c) $O(n^{11})$

d) $O(2^n)$

(See Section 21.2.11 for the solution and discussion.)

Meanwhile:

Quiz 21.3

Suppose $k = 10$. What is the running time of a straight-forward implementation of exhaustive search, as a function of n? (Choose the strongest correct statement.)

a) $O(n^{10})$

b) $O(n^{11})$

c) $O(2^n)$

d) $O(n!)$

(See Section 21.2.11 for the solution and discussion.)

Uh-oh. . . a dynamic programming algorithm that uses the subproblems on page 115 can't beat exhaustive search! And any algorithm with a running time like $O(n^{10})$ is practically useless except in the smallest of graphs. We need another idea.

21.2.4   Color Coding

Why are we using so many subproblems? Oh right, we're keeping track of the vertices S visited by a path to avoid inadvertently creating a path that visits a vertex more than once (recall Quiz 21.1 and Lemma 21.1). Can we get away with tracking less information about a path? Here's the inspired idea, given a graph $G = (V, E)$ and path length bound k:[14]

Color Coding

1. Partition the vertex set V into k groups $V_1, V_2, \ldots, V_k$ so that there is a minimum-cost k-path of G that has exactly one vertex in each group.

2. Among all paths with exactly one vertex in each group, compute one with the minimum cost.

This technique is called *color coding* because if we associate each integer of $\{1, 2, \ldots, k\}$ with a color, we can visualize the ith group V_i of the partition in the first step as the vertices colored i. The second step then seeks a minimum-cost *panchromatic path*, meaning a path in which each color is represented exactly once:

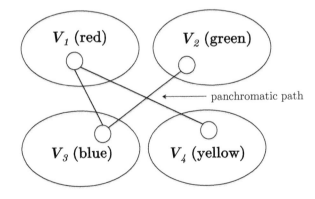

Because there are k colors, a panchromatic path must be a k-path. The converse is false, as a k-path might use some color more than once (and some other color not at all):

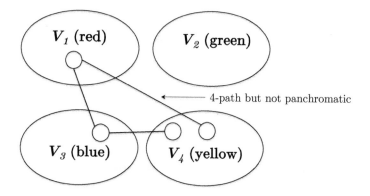

4-path but not panchromatic

If the color-coding plan can be carried out, it would solve the minimum-cost k-path problem: The second step computes a minimum-cost panchromatic path, and the first step ensures that this is also a minimum-cost k-path in G (panchromatic or otherwise).

Skeptical? Perfectly understandable. Why should computing a minimum-cost panchromatic path be any easier than the original problem? And how on earth can we implement the first step without knowing anything about what the minimum-cost k-paths are?

21.2.5   Computing a Minimum-Cost Panchromatic Path

Restricting attention to panchromatic paths simplifies the minimum-cost k-path problem because it frees a dynamic programming algorithm to guard against repeated *colors* instead of repeated vertices. (A repeated vertex implies a repeated color but not vice versa.) Subproblems can track the colors represented in a path, along with an ending vertex, in lieu of the vertices themselves. Why is this a win? Because there are only 2^k subsets of colors, as opposed to the $\Omega(n^k)$ subsets of at most k vertices (see the solution to Quiz 21.2).

Subproblems and Recurrence

For a color subset $S \subseteq \{1, 2, \ldots, k\}$, an *S-path* is a (cycle-free) path with $|S|$ vertices and all colors of S represented. (Panchromatic paths are precisely the S-paths with $S = \{1, 2, \ldots, k\}$.) For a graph $G =$

(V, E) with edge costs and a color assignment (in $\{1, 2, \ldots, k\}$) to each vertex $v \in V$, the subproblems are then:

Minimum-Cost Panchromatic Path: Subproblems

Compute $C_{S,v}$, the minimum cost of an S-path that ends at the vertex $v \in V$ (or $+\infty$, if there is no such path).

(For each non-empty subset $S \subseteq \{1, 2, \ldots, k\}$ of colors, and each vertex $v \in V$.)

The minimum cost of a panchromatic path is the smallest of the solutions to the biggest subproblems (with $S = \{1, 2, \ldots, k\}$); if the graph has no panchromatic paths, all such subproblem solutions will be $+\infty$.

An optimal path P for a subproblem with color subset S and ending vertex v must be built up from an optimal path for a smaller subproblem:

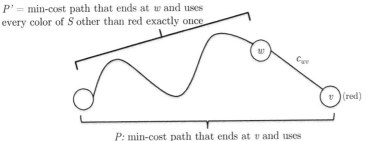

$P' = $ min-cost path that ends at w and uses every color of S other than red exactly once

w

c_{wv}

v (red)

P: min-cost path that ends at v and uses every color of S exactly once

If P's final hop is (w, v), its prefix $P' = P - \{(w, v)\}$ must be a minimum-cost $(S - \{\sigma(v)\})$-path ending at w, where $\sigma(v)$ denotes v's color.[15] This optimal substructure leads immediately to the following recurrence for solving all the subproblems.

Lemma 21.4 (Min-Cost Panchromatic Path Recurrence)
Continuing with the same notation, for every subset $S \subseteq \{1, 2, \ldots, k\}$ of at least two colors and vertex $v \in V$:

$$C_{S,v} = \min_{(w,v) \in E} \left(C_{S-\{\sigma(v)\},w} + c_{wv} \right). \tag{21.7}$$

---

[15]The formal proof of this statement is almost the same as the proof of Lemma 21.1 (see page 113).

Dynamic Programming Algorithm

In turn, this recurrence leads immediately to a dynamic programming algorithm for computing the minimum cost of a panchromatic path (or $+\infty$, if no such path exists):

PanchromaticPath

Input: undirected graph $G = (V, E)$, a real-valued cost c_{vw} for each edge $(v, w) \in E$, and a color $\sigma(v) \in \{1, 2, \ldots, k\}$ for each vertex $v \in V$.
Output: the minimum total cost of a panchromatic path of G (or $+\infty$, if no such path exists).

```
// subproblems (indexed by S ⊆ {1,...,k}, v ∈ V)
```
$A := (2^k - 1) \times |V|$ two-dimensional array
```
// base cases (|S| = 1)
```
for $i = 1$ to k **do**
    for $v \in V$ **do**
        if $\sigma(v) = i$ **then**
            $A[\{i\}][v] := 0$ `// via the empty path`
        else
            $A[\{i\}][v] := +\infty$ `// no such path`
```
// systematically solve all subproblems
```
for $s = 2$ to k **do**   `// s=subproblem size`
    for S with $|S| = s$ **do**
        for $v \in V$ **do**
             `// use recurrence from Lemma 21.4`
            $A[S][v] := \min_{(w,v) \in E} (A[S - \{\sigma(v)\}][w] + c_{wv})$
```
// best of solutions to largest subproblems
```
return $\min_{v \in V} A[\{1, 2, \ldots, k\}][v]$

See Problem 21.5 for an example of the algorithm in action.

21.2.6   Correctness and Running Time

The correctness of the PanchromaticPath algorithm follows by induction (on the subproblem size), with the recurrence in Lemma 21.4

justifying the inductive step. With a little extra bookkeeping, a minimum-cost panchromatic path can be reconstructed in an $O(k)$-time postprocessing step (Problem 21.7).

The running time analysis echoes that of the Bellman-Ford algorithm (see Chapter 18 in *Part 3*). Almost all the work performed by the algorithm occurs in its triple-for loop. Assuming the input graph is represented using adjacency lists, an innermost loop iteration that computes the value of $A[S][v]$ takes $O(\deg(v))$ time, where $\deg(v)$ denotes the degree (the number of incident edges) of the vertex v.[16] For each subset S, the combined time spent solving the associated $|V|$ subproblems is $O(\sum_{v \in V} \deg(v)) = O(m)$, where $m = |E|$ denotes the number of edges.[17] The number of different color subsets S is less than 2^k, so the overall running time of the algorithm is $O(2^k m)$.

Theorem 21.5 (Properties of PanchromaticPath)   *For every graph G with m edges, real-valued edge costs, and assignment of each vertex to a color in $\{1, 2, \ldots, k\}$, the* PanchromaticPath *algorithm runs in $O(2^k m)$ time and returns the minimum cost of a panchromatic path (if one exists) or $+\infty$ (otherwise).*

With running time scaling with 2^k rather than n^k, the PanchromaticPath algorithm improves dramatically over exhaustive search. But the problem we truly care about is the minimum-cost k-path problem without any panchromatic constraint. How does this algorithm help?

21.2.7   Randomization to the Rescue

The first step of the color-coding approach colors the vertices of the input graph so that at least one minimum-cost k-path turns panchromatic. How can we accomplish this without knowing which k-paths are the minimum-cost ones? Time to bring out another tool from our algorithmic toolbox: *randomization*. The hope is that a uniformly random coloring has a healthy chance of rendering some minimum-cost k-path panchromatic, in which case the PanchromaticPath algorithm will find one.

---

[16]Technically, this analysis assumes that every vertex has degree at least 1; degree-0 vertices can be discarded harmlessly in a preprocessing step.

[17]The sum $\sum_{v \in V} \deg(v)$ of vertex degrees is twice the number of edges, with each edge contributing 1 to the degree of each of its two endpoints.

Quiz 21.4

Suppose we assign each vertex of a graph G a color from $\{1, 2, \ldots, k\}$ independently and uniformly at random. Consider a k-path P in G. What is the probability that P winds up panchromatic?

  a) $\frac{1}{k}$

  b) $\frac{1}{k^2}$

  c) $\frac{1}{k!}$

  d) $\frac{k!}{k^k}$

  e) $\frac{1}{k^k}$

(See Section 21.2.11 for the solution and discussion.)

Is this probability of getting lucky—call it p—big or small? Remember that we know an extremely accurate estimate of the factorial function (Stirling's approximation on page 106). Plugging in the approximation in (21.1), with k playing the role of n:

$$p = \frac{k!}{k^k} \approx \frac{1}{k^k} \cdot \sqrt{2\pi k} \left(\frac{k}{e}\right)^k = \frac{\sqrt{2\pi k}}{e^k}. \qquad (21.8)$$

Looks bad—the probability of success (that is, of turning a minimum-cost k-path panchromatic) is less than 1% already when $k = 7$. But if we experiment with a large number of independent random colorings—running the `PanchromaticPath` algorithm for each and remembering the least costly k-path found—*one* lucky coloring is all we need. How many random trials T do we need to ensure a 99% chance that one of our colorings renders a minimum-cost k-path panchromatic?

A trial succeeds with probability p, so it fails with probability $1 - p$. Because the trials are independent, their failure probabilities multiply. The probability that all T trials fail is then $(1 - p)^T$.[18] Remembering that we can bound $1 - p$ from above by e^{-p}:

---

[18] For background on discrete probability, see Appendix B of *Part 1* or the resources at www.algorithmsilluminated.org.

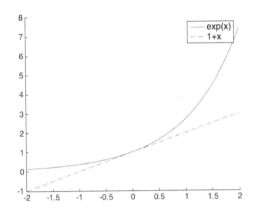

this failure probability is

$$(1 - p)^T \le (e^{-p})^T = e^{-pT}. \tag{21.9}$$

Setting the right-hand side of (21.9) to a target δ (like .01), taking logarithms of both sides, and solving for T shows that

$$T \ge \frac{1}{p} \cdot \ln\left(\frac{1}{\delta}\right) \tag{21.10}$$

independent trials are enough to drive the failure probability down to δ. As expected, the lower the single-trial success probability or the desired failure probability, the greater the number of trials required.

Substituting our success probability p from (21.8) into (21.10) shows that:

Lemma 21.6 (Random Colorings Are Good Enough) *For every graph G, k-path P of G, and failure probability $\delta \in (0, 1)$: If*

$$T \ge \frac{e^k}{\sqrt{2\pi k}} \cdot \ln\left(\frac{1}{\delta}\right),$$

the probability that at least one of T uniformly random colorings turns P panchromatic is at least $1 - \delta$.

The exponential number of trials in Lemma 21.6 may look extravagant, but it's in the same ballpark as the time already required by a single invocation of the PanchromaticPath subroutine. When k is small relative to n—the regime relevant to the motivating application (Section 21.2.1)—the number of trials is much less than the time required to solve the minimum-cost k-path problem on n-vertex graphs by exhaustive search (which, by Quiz 21.3, scales with n^k).

21.2.8   The Final Algorithm

We now have all our ducks in a row: Lemma 21.6 promises that many independent random colorings suffice to implement the first step of the color-coding approach (page 117) and the `PanchromaticPath` algorithm takes care of the second step.

ColorCoding

Input: undirected graph $G = (V, E)$, a real-valued cost c_{vw} for each edge $(v, w) \in E$, a path length k, and a failure probability $\delta \in (0, 1)$.
Output: the minimum total cost of a k-path of G (or $+\infty$, if no such path exists), except with a failure probability of at most δ.

$C_{best} := +\infty$ `// cheapest k-path found so far`
`// number of random trials (from Lemma 21.6)`
$T := (e^k / \sqrt{2\pi k}) \ln \frac{1}{\delta}$ `// round up to an integer`
for $t = 1$ **to** T **do**   `// independent trials`
    for each $v \in V$ **do**   `// choose random coloring`
        $\sigma_t(v) :=$ random number from $\{1, 2, \ldots, k\}$
    `// best panchromatic path for this coloring`
    $C :=$ `PanchromaticPath`(G, c, σ_t) `// see page 120`
    if $C < C_{best}$ **then**   `// found a cheaper k-path!`
        $C_{best} := C$
return C_{best}

21.2.9   Running Time and Correctness

The running time of the `ColorCoding` algorithm is dominated by its $T = O((e^k / \sqrt{k}) \ln \frac{1}{\delta})$ calls to the $O(2^k m)$-time `PanchromaticPath` subroutine, where m denotes the number of edges (Theorem 21.5).

To argue correctness, consider a minimum-cost k-path P^* of G, with total cost C^*.[19] For every coloring σ, the minimum cost of a panchromatic path—the output of the `PanchromaticPath` subroutine— is at least C^*. Whenever σ turns P^* panchromatic, this cost is

---

[19]If G has no k-paths, every invocation of `PanchromaticPath` and the `ColorCoding` algorithm will return $+\infty$ (which is the correct answer).

exactly C^*. By Lemma 21.6, with probability at least $1 - \delta$, at least one of the iterations of the outer loop chooses such a coloring; in this event, the ColorCoding algorithm returns C^*, which is the correct answer.[20]

To summarize:

Theorem 21.7 (Properties of ColorCoding) *For every graph G with n vertices and m edges, real-valued edge costs, path length $k \in \{1, 2, \ldots, n\}$, and failure probability $\delta \in (0, 1)$, the ColorCoding algorithm runs in*

$$O\left(\frac{(2e)^k}{\sqrt{k}} m \ln\left(\frac{1}{\delta}\right)\right) \tag{21.11}$$

time and, with probability at least $1 - \delta$, returns the minimum cost of a k-path of G (if one exists) or $+\infty$ (otherwise).

How should we feel about the running time of the ColorCoding algorithm? The bad news is that the running time bound in (21.11) is exponential—no surprise, given that the minimum-cost k-path problem is NP-hard in general. The good news is that its exponential dependence is confined entirely to the path length k, with only linear dependence on the graph size. In fact, in the special case in which $k \leq c \ln m$ for a constant $c > 0$, the ColorCoding algorithm solves the minimum-cost k-path problem in polynomial time![21,22]

---

[20]The most famous randomized algorithm, QuickSort, has a random running time (ranging from near-linear to quadratic) but is always correct; see Chapter 5 of *Part 1*. The ColorCoding algorithm is the opposite: The outcomes of its coin flips determine whether it's correct but have little effect on its running time.

[21]Observe that $(2e)^{c \ln m} = m^{c \ln(2e)} \approx m^{1.693c}$, which is polynomial in the graph size.

[22]The ColorCoding algorithm is an example of a *fixed-parameter* algorithm, meaning one with a running time of the form $O(f(k) \cdot n^d)$, where n denotes the input size, d is a constant (independent of k and n), and k is a parameter measuring the "difficulty" of the instance. The function f must be independent of n but can have arbitrary dependence (typically exponential or worse) on the parameter k. A fixed-parameter algorithm runs in polynomial time for all instances in which k is sufficiently small relative to n.

The 21st century has seen tremendous progress in understanding which NP-hard problems and parameter choices allow for fixed-parameter algorithms. For a deep dive, check out the book *Parameterized Algorithms*, by Marek Cygan, Fedor V. Fomin, Łukasz Kowalik, Daniel Lokshtanov, Dániel Marx, Marcin Pilipczuk, Michał Pilipczuk, and Saket Saurabh (Springer, 2015).

21.2.10   Revisiting PPI Networks

Fancy guarantees like Theorem 21.7 are nice and all, but how well does the ColorCoding algorithm actually work in the motivating application of finding long linear pathways in PPI networks? The algorithm is a great fit for the application, in which typical path lengths k are in the 10–20 range. (Significantly longer paths, if they exist, are challenging to interpret.) Already with circa-2007 computers, optimized implementations of the ColorCoding algorithm could find linear pathways of length $k = 20$ in major PPI networks with thousands of vertices. This was a significant advance over exhaustive search (which is useless even for $k = 5$) and the competing algorithms at that time (which failed to go much beyond $k = 10$).[23]

21.2.11   Solutions to Quizzes 21.2–21.4

Solution to Quiz 21.2

Correct answer: (b). The number of non-empty subsets $S \subseteq V$ of size at most 10 is the sum $\sum_{i=1}^{10} \binom{n}{i}$ of ten binomial coefficients. Bounding the ith summand from above by n^i and using the formula (20.9) for a geometric series shows that this sum is $O(n^{10})$. Expanding the last binomial coefficient shows that it alone is $\Omega(n^{10})$.[24] With at most ten choices for the endpoint v for each set S, the total number of subproblems is $\Theta(n^{10})$.

Solution to Quiz 21.3

Correct answer: (a). Exhaustive search enumerates the $n \cdot (n-1) \cdot (n-2) \cdots (n-9) = \Theta(n^{10})$ ordered 10-tuples of distinct vertices, computes the cost of each tuple that corresponds to a path (in $O(1)$ time, assuming access to an adjacency matrix populated by edges' costs), and remembers the best of the 10-paths it encounters. The running time of this algorithm is $\Theta(n^{10})$.

---

[23]For more details, check out the paper "Algorithm Engineering for Color-Coding with Applications to Signaling Pathway Detection," by Falk Hüffner, Sebastian Wernicke, and Thomas Zichner (*Algorithmica*, 2008).

[24]Recall that big-omega notation is analogous to "greater than or equal to." Formally, $f(n) = \Omega(g(n))$ if and only if there is a constant $c > 0$ such that $f(n) \geq c \cdot g(n)$ for all sufficiently large n. Also, $f(n) = \Theta(g(n))$ if and only if $f(n) = O(g(n))$ and $f(n) = \Omega(g(n))$.

Solution to Quiz 21.4

Correct answer: (d). There are k^k different ways to color the vertices of P (with k color choices for each of the k vertices), each equally likely (with probability $1/k^k$ each). How many of these render P panchromatic? There are k choices for the vertex receiving the color 1, then $k - 1$ remaining choices for the vertex receiving the color 2, and so on, for a total of $k!$ panchromatic colorings. The probability of panchromaticity is, therefore, $k!/k^k$.

21.3   Problem-Specific Algorithms vs. Magic Boxes

21.3.1   Reductions and Magic Boxes

Bespoke solutions to fundamental problems like the `BellmanHeldKarp` (Section 21.1.6) and `ColorCoding` (Section 21.2.8) algorithms are deeply satisfying. But before investing the effort necessary to design or code up a new algorithm, you should always ask yourself:

> Is this problem a special case or a thinly disguised version of one that I already know how to solve?

If the answer is "no," or even "yes, but the algorithms for the more general problem aren't good enough for this application," you've justified proceeding to problem-specific algorithm development.

Throughout this book series, we've seen several problems for which the answer is "yes." For example, median-finding reduces to sorting, all-pairs shortest paths reduces to single-source shortest paths, and the longest common subsequence problem is a special case of the sequence alignment problem (Section 19.5.2). Such reductions transfer tractability from one problem B to another problem A:

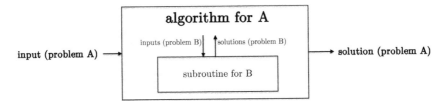

Our reductions thus far have been to problems B for which we ourselves had already designed a fast algorithm. But a reduction from a problem A to a problem B retains its power *even if you don't personally know* how to solve B efficiently. As long as someone gives you a magic box (such as an inscrutable piece of software) that solves problem B, you're good to go: To solve problem A, execute the reduction to problem B and invoke the magic box as needed.

21.3.2   MIP and SAT Solvers

A "magic box" probably sounds like pure fantasy, akin to a unicorn or the fountain of youth. Could one really exist? Sections 21.4 and 21.5 describe two of the closest approximations out there—solvers for mixed integer programming (MIP) and satisfiability (SAT) problems. By a "solver," we mean a sophisticated algorithm that has been carefully tuned and expertly implemented as ready-to-use software. MIP and SAT are very general problems, expressive enough to capture most of the problems studied in this book series as special cases.

Several decades worth of engineering effort and ingenuity have been poured into state-of-the-art MIP and SAT solvers. For this reason, despite their generality, such solvers semi-reliably solve medium-size instances of NP-hard problems in a tolerable amount of time. Solver performance varies widely with the problem (and many other factors) but, as a rough guideline, you can reasonably cross your fingers and hope for inputs with size in the thousands to be solved in less than a day, and often much faster. In some applications, MIP and SAT solvers are unreasonably effective even for large instances, with input sizes in the millions.

21.3.3   What You Will and Won't Learn

The goals of Sections 21.4 and 21.5 are modest. They do not explain how MIP and SAT solvers work—that would require a whole other book. Instead, they prepare you to be an educated client of these solvers.[25]

---

[25]From 30,000 feet, the basic idea is: Recursively search the space of candidate solutions à la depth-first search, applying the clues gleaned so far to aggressively prune not-yet-examined candidates (such as candidates that cannot possibly have objective function value better than the best solution already discovered), backtracking as needed. The hope is that most of the search space gets pruned without explicit examination. To explore these ideas further, look up "branch and bound" (for MIP solvers) and "conflict-driven clause learning" (for SAT solvers).

> ### Goals of Sections 21.4–21.5
>
> 1. Be aware that semi-reliable magic boxes called MIP and SAT solvers exist and can be unreasonably effective at tackling NP-hard problems in practice. (Not enough programmers know this!)
>
> 2. See examples of encodings of NP-hard problems as MIP and SAT problems.
>
> 3. Know where to go next to learn more.

21.3.4   A Rookie Mistake Revisited

MIP and SAT solvers routinely crack tough problems, but don't get fooled into thinking that NP-hardness doesn't matter in practice (the third rookie mistake in Section 19.6). When applying such a solver to an NP-hard problem, keep your fingers crossed and have at the ready a plan B (like a fast heuristic algorithm) in case the solver fails. And make no mistake: There will be some instances out there, including fairly small ones, that can bring your solver to its knees. You take whatever you can with NP-hard problems, and semi-reliable magic boxes are about as good as it gets.

21.4   Mixed Integer Programming Solvers

Most discrete optimization problems can be cast as mixed integer programming (MIP) problems.[26] Whenever you're faced with an NP-hard optimization problem that you can encode efficiently as a MIP problem, throwing the latest and greatest MIP solver at it is probably worth a shot.

21.4.1   Example: The Knapsack Problem

In the knapsack problem (Section 19.4.2), the input is specified by $2n + 1$ positive integers: n item values $v_1, v_2, \ldots, v_n$, n item sizes $s_1, s_2, \ldots, s_n$, and a knapsack capacity C. For example:

---

[26]This is the same anachronistic use of the word "programming" as in dynamic programming (or television programming); it refers to planning, not coding.

	Value	Size
Item #1	6	5
Item #2	5	4
Item #3	4	3
Item #4	3	2
Item #5	2	1
Knapsack capacity: 10		

The goal is to compute a subset of the items with the maximum-possible total value, subject to having a total size of at most the knapsack capacity. The problem specification thus spells out three things:

1. *The decisions to be made:* for each of the n items, whether to include it in the subset. One convenient way to encode these per-item binary decisions numerically is with 0-1 variables, called *decision variables:*

$$x_j = \begin{cases} 1 & \text{if item } j \text{ is included} \\ 0 & \text{if item } j \text{ is excluded} \end{cases}. \qquad (21.12)$$

2. *The constraints to be respected:* the sum of the sizes of the chosen items should be at most the knapsack capacity C. This constraint is easily expressed in terms of the decision variables, with item j contributing s_j to the total size if it's included (with $x_j = 1$) and 0 if it's excluded (with $x_j = 0$):

$$\underbrace{\sum_{j=1}^{n} s_j x_j}_{\substack{\text{total size of} \\ \text{chosen subset}}} \leq C. \qquad (21.13)$$

3. *The objective function:* the sum of the values of the chosen items should be as large as possible (subject to the capacity constraint). This objective function is equally easy to express (with j contributing value v_j if included and 0 if excluded):

$$\text{maximize} \underbrace{\sum_{j=1}^{n} v_j x_j}_{\substack{\text{total value of} \\ \text{chosen subset}}}. \qquad (21.14)$$

Guess what? In (21.12)–(21.14), you've just seen your first example of an *integer program*. For example, in the 5-item instance described above, this integer program reads:

$$\text{maximize} \quad 6x_1 + 5x_2 + 4x_3 + 3x_4 + 2x_5 \qquad (21.15)$$

$$\text{subject to} \quad 5x_1 + 4x_2 + 3x_3 + 2x_4 + x_5 \leq 10 \qquad (21.16)$$

$$x_1, x_2, x_3, x_4, x_5 \in \{0, 1\}. \qquad (21.17)$$

This is exactly the sort of description that can be fed directly into a magic box called a *mixed integer programming (MIP) solver*.[27] For example, to solve the integer program in (21.15)–(21.17) using Gurobi Optimizer, the leading commercial MIP solver, you literally just call it from the command line with the following input file:

```
Maximize 6 x(1) + 5 x(2) + 4 x(3) + 3 x(4) + 2 x(5)
subject to
5 x(1) + 4 x(2) + 3 x(3) + 2 x(4) + x(5) <= 10
binary
x(1) x(2) x(3) x(4) x(5)
end
```

Magically, the solver then spits out the optimal solution (in this case, with $x_1 = 0$, $x_2 = x_3 = x_4 = x_5 = 1$, and objective function value 14).[28]

21.4.2   MIPs More Generally

In general, a MIP is specified by the three ingredients listed in Section 21.4.1: the decision variables, along with the values they can

---

[27]Why "mixed"? Because these solvers also accommodate decision variables that can take on real (not necessarily integer) values. Some authors refer to MIPs as integer linear programs (ILPs) or simply integer programs (IPs). Others reserve the latter term for MIPs in which all the decision variables are integer-valued.

A MIP in which none of the decision variables are required to be integers is called a *linear program (LP)*. State-of-the-art solvers work particularly well for LPs, and often solve thousands of them in the course of solving a single MIP. (Relatedly, linear programming is a polynomial-time solvable problem while general mixed integer programming is an NP-hard problem.)

[28]For this toy example, the input file is easy enough to create by hand. For larger instances, you'll want to write a program that generates the input file automatically or interacts directly with the solver API.

assume (such as 0 or 1, or any integer, or any real number); the constraints; and the objective function. The one important restriction is that both the constraints and the objective function should be *linear* in the decision variables.[29] In other words, it's OK to scale a decision variable by a constant, and it's OK to add decision variables together, but that's it. For example, in (21.15)–(21.17), you don't see any terms like x_j^2, $x_j x_k$, $1/x_j$, e^{x_j}, and so on.[30]

Problem: Mixed Integer Programming

Input: A list of (binary, integer, or real-valued) decision variables $x_1, x_2, \ldots, x_n$; a linear objective function to be maximized or minimized, specified by its coefficients $c_1, c_2, \ldots, c_n$; and m linear constraints, with each constraint i specified by its coefficients $a_{i1}, a_{i2}, \ldots, a_{in}$ and right-hand side b_i.

Output: An assignment of values to $x_1, x_2, \ldots, x_n$ that optimizes the objective function ($\sum_{j=1}^{n} c_j x_j$) subject to the m constraints ($\sum_{j=1}^{n} a_{ij} x_j \leq b_i$ for all $i = 1, 2, \ldots, m$). (Or if no assignment satisfies all the constraints, report this fact.)

Even with the linearity restriction, it's often embarrassingly simple to express NP-hard optimization problems as MIPs. For example, consider the *two-dimensional* knapsack problem, where every item j now has a *weight* w_j in addition to a value v_j and size s_j; in addition to the knapsack capacity C, there is a weight bound W. The goal is then to choose the maximum-value subset of items with total size at most C and total weight at most W. As a graduate of the *Algorithms Illuminated* dynamic programming boot camp, you could knock out an algorithm for this problem without much trouble. But you couldn't do it as quickly as you could add the constraint

$$\sum_{j=1}^{n} w_j x_j \leq W \tag{21.18}$$

---

[29]Thus, "MILP" (for mixed integer linear program) would be more precise than "MIP," though also less pleasing to the ear...

[30]State-of-the-art solvers can also accommodate limited types of nonlinearity (like quadratic terms) but are typically much faster with linear constraints and objective functions.

to the knapsack MIP (21.12)–(21.14)!

Familiar optimization problems like the maximum-weight independent set (Section 19.4.2), minimum makespan (Section 20.1), and maximum coverage (Section 20.2) problems are almost equally easy to express as MIPs (see Problem 21.9).[31] MIPs are also the basis for the state-of-the-art exact algorithms for the TSP (Section 19.1.2), although this application is much more sophisticated (see Problem 21.10).[32]

A problem can generally be formulated as a MIP problem in several different ways, with some formulations leading to better solver performance than others (in some cases by orders of magnitude). If your first attempt at tackling an optimization problem with a MIP solver fails, consider experimenting with alternative encodings. As with algorithms, the design of good MIP formulations takes practice; the resources in footnote 32 will get you started.

Finally, if your MIP solver is taking too long to complete no matter which formulation you try, you can interrupt it after a target amount of time and use the best feasible solution found up to that point. (MIP solvers typically generate a sequence of successively better feasible solutions, analogous to the local search algorithms of Sections 20.4–20.5, en route to an optimal solution.) Stopping early effectively turns a MIP solver into a fast heuristic algorithm.

21.4.3   MIP Solvers: Some Starting Points

Now that you're amped up to apply a MIP solver to your favorite problem, where should you start? As of this writing (in 2020), there is a huge gulf in performance between commercial and non-commercial MIP solvers. Currently, Gurobi Optimizer is generally viewed as the fastest and most robust MIP solver, with runners-up including CPLEX and FICO Xpress. University students and staff can obtain

---

[31]For starters, a constraint of the form $\sum_{j=1}^{n} a_{ij}x_j \geq b_i$ can be represented by the equivalent constraint $\sum_{j=1}^{n} (-a_{ij})x_j \leq -b_i$, and an equality constraint $\sum_{j=1}^{n} a_{ij}x_j = b_i$ can be represented by a pair of inequality constraints.

[32]For many more examples and tricks of the trade, check out the (free) documentation for the solvers listed in Section 21.4.3, or the textbook *Model Building in Mathematical Programming*, by H. Paul Williams (*Wiley*, 5th edition, 2013). The examples in Dan Gusfield's book *Integer Linear Programming in Computational and Systems Biology* (Cambridge, 2019) slant toward biological applications but are useful broadly for MIP (and especially Gurobi Optimizer) newbies.

free academic licenses for these solvers (for research and educational purposes only).

If you're stuck using a non-commercial solver, good starting points include SCIP, CBC, MIPCL, and GLPK. CBC and MIPCL have more liberal licensing agreements than the other two, which are free for non-commercial use only.

You can decouple the tasks of formulating a MIP for your problem and describing that MIP to a particular solver by specifying your MIP in a high-level solver-independent modeling language, such as the Python-based CVXPY. You can then experiment easily with all the solvers supported by that language, with your high-level specification automatically compiled into the format expected by the solver.

21.5 Satisfiability Solvers

In many applications, the primary goal is to figure out whether a feasible solution exists (and, if so, to find some such solution) rather than to optimize a numerical objective function. Problems of this type can often be cast as satisfiability (SAT) problems. Whenever you're faced with an NP-hard problem that you can encode efficiently as a SAT problem, throwing the latest and greatest SAT solver at it is probably worth a shot.

21.5.1 Example: Graph Coloring

One of the oldest graph problems out there, studied extensively already in the 19th century, is the *graph coloring* problem. A *k-coloring* of an undirected graph $G = (V, E)$ is an assignment $\sigma(v)$ of each of its vertices $v \in V$ to a color in $\{1, 2, \ldots, k\}$ such that no edge is monochromatic (that is, $\sigma(v) \neq \sigma(w)$ whenever $(v, w) \in E$).[33] A graph with a *k*-coloring is called—wait for it—*k-colorable.*[34] For

---

[33]The `ColorCoding` algorithm (Section 21.2.8) uses random colorings—which are generally not *k*-colorings—internally as a device to achieve a faster running time. In this section, the problem studied explicitly concerns *k*-colorings.

[34]The most famous result in all of graph theory is the "Four Color Theorem," stating that every planar graph—a graph that can be drawn on a piece of paper without any edge crossings—is 4-colorable. (The second graph in this section shows that four colors may be necessary.) Equivalently, as it turns out, maps need only four colors of ink to ensure that every pair of neighboring countries can be colored with different colors (assuming each country is a contiguous region).

example, a wheel graph with six spokes is 3-colorable, while a wheel graph with five spokes is not (as you should check):

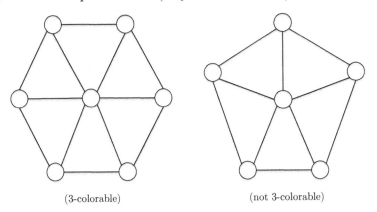

(3-colorable)   (not 3-colorable)

Problem: Graph Coloring

Input: An undirected graph $G = (V, E)$ and a positive integer k.

Output: A k-coloring of G, or a correct declaration that G is not k-colorable.

The graph coloring problem is not purely recreational. For example, the problem of assigning classes to one of k classrooms is exactly a graph coloring problem (with one vertex per class and an edge between each pair of classes that overlap in time). For a high-stakes application of graph coloring-type problems, see Chapter 24.

21.5.2   Satisfiability

The non-numerical and rule-based nature of the graph coloring problem suggests expressing decision variables and constraints using the formalism of *logic* rather than arithmetic. Instead of numerical decision variables, we'll use *Boolean* variables, which can take on only the values true and false. A *truth assignment* specifies one of these two values for each variable. Constraints, which are also called *clauses*, are then logical formulas that express restrictions on the permitted truth assignments. A seemingly simple type of constraint, called a *disjunction of literals*, uses only the logical "or" (denoted by ∨) and logical "not" (denoted by $\neg$) operations.[35] For example, the constraint $x_1 \vee \neg x_2 \vee x_3$ is a disjunction of literals and is satisfied unless you screw up all three of its assignment requests (by setting x_1 and x_3 to false and x_2 to true):

Value of x_1	Value of x_2	Value of x_3	$x_1 \vee \neg x_2 \vee x_3$ Satisfied?
true	true	true	yes
true	true	false	yes
true	false	true	yes
false	true	true	yes
true	false	false	yes
false	true	false	no
false	false	true	yes
false	false	false	yes

In general, disjunctions of literals are easygoing creatures: one with k literals, each corresponding to a distinct decision variable, forbids one and only one of the 2^k ways to assign values to its variables.

An instance of satisfiability (SAT) is specified by its variables (restricted to be Boolean) and constraints (restricted to be disjunctions of literals).

Problem: Satisfiability

Input: A list of Boolean decision variables $x_1, x_2, \ldots, x_n$; and a list of constraints, each a disjunction of one or more literals.

Output: A truth assignment to $x_1, x_2, \ldots, x_n$ that satisfies every constraint, or a correct declaration that no such truth assignment exists.

21.5.3   Encoding Graph Coloring as SAT

Is the SAT problem, with its mere Boolean variables and disjunctions of literals, expressive enough to encode other interesting problems? For example, in the graph coloring problem, we'd ideally like one

---

[35]"Literal" means a decision variable x_i or its negation $\neg x_i$, and "disjunction" means logical "or."

(non-Boolean) decision variable per vertex, each taking on one of k different values (one per possible color).

With a little practice, you can encode a surprisingly large number of problems as SAT.[36],[37] For example, to encode an instance of graph coloring specified by the graph $G = (V, E)$ and integer k, we can use k variables per vertex; for each vertex $v \in V$ and color $i \in \{1, 2, \ldots, k\}$, the Boolean variable x_{vi} indicates whether vertex v is assigned the color i.

What about the constraints? For an edge $(v, w) \in E$ and color i, the constraint

$$\neg x_{vi} \vee \neg x_{wi} \qquad (21.19)$$

is not satisfied precisely when both v and w are colored i. In tandem, the $|E| \cdot k$ constraints of the form (21.19) enforce that no edge is monochromatic.

We're not quite done, as all the constraints of the form (21.19) are satisfied by the all-false truth assignment (corresponding to no vertex receiving any color). But we can add one constraint

$$x_{v1} \vee x_{v2} \vee \cdots \vee x_{vk} \qquad (21.20)$$

for each vertex $v \in V$ that is not satisfied precisely when v receives no color. Every k-coloring of G translates to a truth assignment that satisfies all the constraints and now, conversely, every truth assignment that satisfies all the constraints encodes one or more k-colorings of G.[38]

The system of constraints defined by (21.19) and (21.20) is exactly the sort of description that can be fed directly into a magic box called

---

[36] For many examples, including classic applications to hardware and software verification, see the *Handbook of Satisfiability*, edited by Armin Biere, Marijn Heule, Hans van Maaren, and Toby Walsh (IOS Press, 2009). Or, if you've been wondering what Donald E. Knuth has been up to lately, check out *Satisfiability*, Fascicle 6 of Volume 4 of *The Art of Computer Programming* (Addison-Wesley, 2015). Another fun fact: SAT solvers were recently employed to break the once-secure cryptographic hash function SHA-1; see "The First Collision for Full SHA-1," by Marc Stevens, Elie Bursztein, Pierre Karpman, Ange Albertini, and Yarik Markov (*Proceedings of the 37th CRYPTO Conference*, 2017).

[37] In fact, the Cook-Levin theorem (Theorems 22.1 and 23.2) shows that SAT is a "universal" problem in a precise sense; see Section 23.6.3.

[38] The constraints (21.20) allow vertices to receive more than one color, but the constraints (21.19) ensure that every way of choosing among the assigned colors produces a k-coloring of G.

a *satisfiability (SAT) solver*. For example, to check whether a complete graph on three vertices is 2-colorable using MiniSAT, a popular open-source SAT solver, you literally just call it from the command line with the following input file:

```
p cnf 6 9
1 4 0
2 5 0
3 6 0
-1 -2 0
-4 -5 0
-1 -3 0
-4 -6 0
-2 -3 0
-5 -6 0
```

Magically, the solver spits out a (correct) declaration that there is no way to satisfy all the constraints.[39]

21.5.4   SAT Solvers: Some Starting Points

As of this writing (in 2020), there are lots of good options for freely available SAT solvers. In fact, at least once every two years, SAT nerds from around the world gather and run Olympic-style competitions (complete with medals) between the latest and greatest solvers, each evaluated across a range of difficult benchmark instances. There are dozens of submissions to each competition, most of which are open-source.[40] If you want just one recommendation, MiniSAT, which combines good performance with ease of use and a permissive license (the MIT license), is a popular choice.[41]

---

[39]The first line of the file warns the solver that the SAT instance has six decision variables and nine constraints; the "cnf" stands for "conjunctive normal form" and indicates that each constraint is a disjunction of literals. Numbers between 1 and 6 refer to variables, with "-" indicating negation. The first three and last three variables correspond to the first color and second color, respectively. The first three and last six constraints are of the form in (21.20) and (21.19), respectively. Zeroes mark the ends of constraints.

[40]See www.satcompetition.org.

[41]And to up your SAT-solving game to the next level, look up "satisfiability modulo theories (SMT)" solvers, such as Microsoft's z3 solver (which is also freely available under the MIT license).

The Upshot

☆ Solving the TSP by exhaustive search requires time scaling with $n!$, where n is the number of vertices.

☆ The Bellman-Held-Karp dynamic programming algorithm solves the TSP in $O(n^2 2^n)$ time.

☆ The key idea in the Bellman-Held-Karp algorithm is to parameterize subproblems by a subset of vertices to be visited exactly once and which of those vertices should be visited last.

☆ In the minimum-cost k-path problem, the input is an undirected graph with real-valued edge costs, and the goal is to compute a cycle-free path visiting k vertices with the minimum-possible sum of edge costs.

☆ Solving the minimum-cost k-path problem by exhaustive search requires time scaling with n^k, where n is the number of vertices.

☆ The color-coding algorithm solves the minimum-cost k-path problem in $O((2e)^k m \ln \frac{1}{\delta})$ time, where m is the number of edges and δ is a user-specified failure probability.

☆ The first key idea in the color-coding algorithm is a dynamic programming subroutine that, given an assignment of one of k colors to each vertex of the input graph, computes in $O(2^k m)$ time a minimum-cost panchromatic path.

☆ The second key idea is to experiment with $O(e^k \ln \frac{1}{\delta})$ independent and uniformly random vertex colorings; with probability at least $1 - \delta$, at least one will render some minimum-cost k-path panchromatic.

☆ A mixed integer program (MIP) is specified by numerical decision variables, linear constraints, and a linear objective function.

☆ Most discrete optimization problems can be formulated as MIP problems.

☆ An instance of satisfiability (SAT) is specified by Boolean decision variables and constraints that are disjunctions of literals.

☆ Most feasibility-checking problems can be formulated as SAT problems.

☆ State-of-the-art MIP and SAT solvers can semi-reliably solve medium-size instances of NP-hard problems.

Test Your Understanding

Problem 21.1 *(S)* Does the `BellmanHeldKarp` algorithm for the TSP (Section 21.1.6) refute the P $\neq$ NP conjecture? (Choose all that apply.)

a) Yes, it does.

b) No. A polynomial-time algorithm for the TSP does not necessarily refute the P $\neq$ NP conjecture.

c) No. Because the algorithm uses an exponential (in the input size) number of subproblems, it does not always run in polynomial time.

d) No. Because the algorithm might perform an exponential amount of work to solve a single subproblem, it does not always run in polynomial time.

e) No. Because the algorithm might perform an exponential amount of work to extract the final solution from the solutions to its subproblems, it does not always run in polynomial time.

Problem 21.2 *(S)* For the TSP input in Quiz 20.7 (page 76), what are the final subproblem array entries of the `BellmanHeldKarp` algorithm from Section 21.1.6?

Problem 21.3 *(S)* Consider the following proposed subproblems for an instance $G = (V, E)$ of the TSP:

> ### Subproblems (Attempt)
>
> Compute $C_{i,v}$, the minimum cost of a cycle-free path that begins at vertex 1, ends at vertex v, and visits exactly i vertices (or $+\infty$, if there is no such path).
>
> (For each $i \in \{2, 3, \ldots, |V|\}$ and $v \in V - \{1\}$.)

What prevents us from using these subproblems, with i as the measure of subproblem size, to design a polynomial-time dynamic programming algorithm for the TSP? (Choose all that apply.)

a) The number of subproblems is super-polynomial in the input size.

b) Optimal solutions to bigger subproblems cannot be computed easily from optimal solutions to smaller subproblems.

c) The optimal tour cannot be computed easily from the optimal solutions to all the subproblems.

d) Nothing!

Problem 21.4 *(S)* Which of the following problems can be solved in $O(n^2 2^n)$ time for n-vertex graphs using a minor variation of the `BellmanHeldKarp` algorithm? (Choose all that apply.)

a) Given an n-vertex undirected graph, determine whether it has a Hamiltonian path (a cycle-free path with $n - 1$ edges).

b) Given an n-vertex directed graph, determine whether it has a directed Hamiltonian path (a cycle-free directed path with $n - 1$ edges).

c) Given a complete undirected graph and real-valued edge costs, compute the maximum cost of a traveling salesman tour.

d) Given a complete n-vertex directed graph (with all $n(n-1)$ directed edges present) and real-valued edge costs, compute the minimum cost of a directed traveling salesman tour (a directed cycle that visits every vertex exactly once).

e) The cycle-free shortest path problem defined on page 29 in Section 19.5.4.

Problem 21.5 *(S)* For the instance

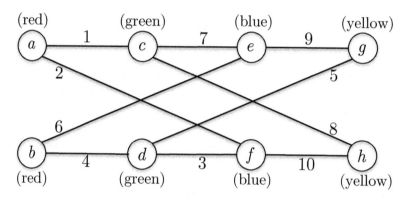

what are the final subproblem array entries of the `PanchromaticPath` algorithm from Section 21.2.5?

Problem 21.6 *(H)* Propose an implementation of a postprocessing step that reconstructs a minimum-cost traveling salesman tour from the subproblem array computed by the `BellmanHeldKarp` algorithm. Can you achieve a linear (in the number of vertices) running time, perhaps after adding some extra bookkeeping to the `BellmanHeldKarp` algorithm?

Problem 21.7 *(H)* Propose an implementation of a postprocessing step that reconstructs a minimum-cost panchromatic path from the subproblem array computed by the `PanchromaticPath` algorithm. Can you achieve a linear (in the number of colors) running time, perhaps after adding some extra bookkeeping to the `PanchromaticPath` algorithm?

Challenge Problems

Problem 21.8 *(H)* Optimize the `BellmanHeldKarp` algorithm for the TSP (Section 21.1.6) so that its memory requirement drops from $O(n \cdot 2^n)$ to $O(\sqrt{n} \cdot 2^n)$ for n-vertex instances. (You are responsible only for computing the minimum cost of a tour, not an optimal tour itself.)

Problem 21.9 *(S)* Show how to encode instances of the following problems as mixed integer programs:

(a) Maximum-weight independent set (Section 19.4.2).

(b) Makespan minimization (Section 20.1.1).

(c) Maximum coverage (Section 20.2.1).

Problem 21.10 *(H)* Given a TSP instance G with vertex set $V = \{1, 2, \ldots, n\}$ and edge costs c, consider the MIP

$$
\begin{array}{lll}
\text{minimize} & \sum_{i=1}^{n} \sum_{j \neq i} c_{ij} x_{ij} & \text{(21.21)} \\
\text{subject to} & \sum_{j \neq i} x_{ij} = 1 & \text{[for every vertex } i] \quad \text{(21.22)} \\
& \sum_{j \neq i} x_{ji} = 1 & \text{[for every vertex } i] \quad \text{(21.23)} \\
& x_{ij} \in \{0, 1\} & \text{[for every } i \neq j]. \quad \text{(21.24)}
\end{array}
$$

The intent is to encode a traveling salesman tour (oriented in one of the two possible directions) with x_{ij} equal to 1 if and only if the tour visits j immediately after i. The constraints (21.22)–(21.23) enforce that each vertex has exactly one immediate predecessor and one immediate successor on the tour.

(a) Prove that, for every TSP instance G and traveling salesman tour of G, there is a feasible solution of the corresponding MIP (21.21)–(21.24) with the same objective function value.

(b) Prove that there is a TSP instance G and a feasible solution of the corresponding MIP (21.21)–(21.24) with objective function value strictly less than the minimum total cost of a traveling salesman tour of G. (Thus, this MIP has spurious feasible solutions, above and beyond the traveling salesman tours, and does not correctly encode the TSP.)

(c) Suppose we throw in the following additional constraints:

$$y_{1j} = (n-1)x_{1j} \qquad \text{[for all } j \in V - \{1\}] \quad (21.25)$$
$$y_{ij} \leq (n-1)x_{ij} \qquad \text{[for all } i \neq j] \qquad\qquad (21.26)$$
$$\textstyle\sum_{j \neq i} y_{ji} - \sum_{j \neq i} y_{ij} = 1 \quad \text{[for all } i \in V - \{1\}] \quad (21.27)$$
$$y_{ij} \in \{0, 1, \ldots, n-1\} \quad \text{[for all } i \neq j], \qquad (21.28)$$

where the y_{ij}'s are additional decision variables.

Reprove (a) for the expanded MIP (21.21)–(21.28).

(d) Prove that, for every TSP instance G, every feasible solution of the corresponding expanded MIP (21.21)–(21.28) translates to a traveling salesman tour of G with the same objective function value. (As a consequence, the expanded MIP correctly encodes the TSP.)[42]

Problem 21.11 *(H)* Show how to encode an instance of the satisfiability problem as a mixed integer program.

Problem 21.12 *(H)* For a positive integer k, the *k-SAT* problem is the special case of the SAT problem in which every constraint has at most k literals. Show that the 2-SAT problem can be solved in $O(m+n)$ time, where m and n denote the number of constraints and variables, respectively. (You can assume that the input is represented as an array of literals and an array of constraints, with pointers from each constraint to its literals and from each literal to the constraints that contain it.)[43]

---

[42] Adding still more constraints, while not necessary for correctness, provides a MIP solver with more clues to work with and can result in significant speedups. For example, adding the (logically redundant) inequalities $x_{ij} + x_{ji} \leq 1$ for all $i \neq j$ to the expanded MIP (21.21)–(21.28) typically reduces the amount of time required to solve it. State-of-the-art MIP solvers that are tailored to the TSP, such as the Concorde TSP solver, draw from an *exponentially large* set of additional inequalities, generated lazily on an as-needed basis. (To learn more, look up the "subtour relaxation" for the TSP.)

[43] The satisfiability formulation in Section 21.5.3 can be viewed as a reduction from the k-coloring problem to the k-SAT problem. Through this formulation, the 2-SAT algorithm in this problem translates to a linear-time algorithm for checking whether a graph is 2-colorable. (A 2-colorable graph is also called "bipartite.") Alternatively, 2-colorability can be checked directly in linear time using breadth-first search.

Problem 21.13 *(H)* Meanwhile, the 3-SAT problem is NP-hard (Theorem 22.1). But can we at least improve over exhaustive search, which enumerates all the 2^n possible truth assignments to the n decision variables? Here's a randomized algorithm, parameterized by a number of trials T:

Schöning

Input: an n-variable instance of 3-SAT and a failure probability $\delta \in (0, 1)$.
Output: with probability at least $1 - \delta$, either a truth assignment that satisfies all the constraints or a correct declaration that none exist.

```
ta := length-n Boolean array      // truth assignment
for t = 1 to T do                 // T independent trials
    for i = 1 to n do // random initial assignment
        ta[i] := "true" or "false"    // 50% chance each
    for k = 1 to n do             // n local modifications
        if ta satisfies all constraints then       // done!
            return ta
        else                      // fix a violated constraint
            choose an arbitrary violated constraint C
            choose variable xi in C, uniformly at random
            ta[i] := ¬ta[i]                // flip its value
return "no solution"              // give up on the search
```

(a) Prove that whenever there is no truth assignment that satisfies all the constraints of the given 3-SAT instance, the Schöning algorithm returns "no solution."

(b) For this and the next three parts, restrict attention to inputs with a satisfying truth assignment (that is, a truth assignment that satisfies all the constraints). Let p denote the probability, over the coin flips of the Schöning algorithm, that an iteration of the outermost for loop discovers a satisfying assignment. Prove that, with $T = \frac{1}{p} \ln \frac{1}{\delta}$ independent random trials, the Schöning algorithm finds a satisfying assignment with probability at least $1 - \delta$.

(c) In this and the next part, let ta^* denote a satisfying assignment of the given 3-SAT instance. Prove that every variable flip made by the Schöning algorithm in its inner loop has at least a 1 in 3 chance of increasing the number of variables with the same value in both ta and ta^*.

(d) Prove that the probability that a uniformly random truth assignment agrees with ta^* on at least $n/2$ variables is at least 50%.

(e) Prove that the probability p defined in (b) is at least $1/(2 \cdot 3^{n/2})$; hence, with $T = 2 \cdot 3^{n/2} \ln \frac{1}{\delta}$ trials, the Schöning algorithm returns a satisfying assignment with probability at least $1 - \delta$.

(f) Conclude that there is a randomized algorithm that solves the 3-SAT problem (with failure probability at most δ) in $O((1.74)^n \ln \frac{1}{\delta})$ time—exponentially faster than exhaustive search.[44]

Programming Problems

Problem 21.14 Implement in your favorite programming language the BellmanHeldKarp algorithm for the TSP (Section 21.1.6). As in Problem 20.15, try out your implementation on instances with edge costs chosen independently and uniformly at random from the set $\{1, 2, \ldots, 100\}$ or, alternatively, for vertices that correspond to points chosen independently and uniformly at random from the unit square

---

[44]This algorithm was proposed by Uwe Schöning. His paper "A Probabilistic Algorithm for k-SAT Based on Limited Local Search and Restart" (*Algorithmica*, 2002) achieves a running time bound of $O((1.34)^n \ln \frac{1}{\delta})$ on n-variable instances through a more careful analysis, and also extends the algorithm and analysis to the k-SAT problem for all k (with the base in the exponential running time increasing from $\approx \frac{4}{3}$ to $\approx 2 - \frac{2}{k}$). Some slightly faster algorithms (both randomized and deterministic) have been developed since, but none have achieved a running time of $O((1.3)^n)$.

Section 23.5 describes the "Exponential Time Hypothesis (ETH)" and "Strong Exponential Time Hypothesis (SETH)," both of which postulate that the flaws of the Schöning algorithm are shared by all k-SAT algorithms. The ETH is a bolder form of the P $\neq$ NP conjecture asserting that solving 3-SAT requires exponential time—time $\Omega(a^n)$ for some constant $a > 1$—and hence the only improvements possible to the Schöning algorithm are to the base of the exponent. The SETH asserts that the base of the exponent of the running time of k-SAT algorithms must degrade to 2 as k grows large.

(with edge costs equal to Euclidean distances). How large an input size (that is, how many vertices) can your program reliably process in under a minute? What about in under an hour? Is the biggest bottleneck time or memory? Does it help if you implement the optimization in Problem 21.8? (See www.algorithmsilluminated.org for test cases and challenge data sets.)

Problem 21.15 Try out one or more MIP solvers on the same types of TSP instances you considered in Problem 21.14, using the MIP formulation in Problem 21.10. How large an input size can the solver reliably process in under a minute, or under an hour? How much does the answer vary with the solver? Does it help if you add the additional inequalities from footnote 42? (See www.algorithmsilluminated. org for test cases and challenge data sets.)

Chapter 22

Proving Problems NP-Hard

Chapters 20 and 21 supplied you with an algorithmic toolbox for tackling NP-hard problems, be it by fast heuristic algorithms or better-than-exhaustive-search exact algorithms. How do you know when you must resort to this toolbox? If your boss hands you a computational problem and tells you it's NP-hard, fine. But what if you're the boss? Problems in the wild don't show up tattooed with their computational status, and recognizing NP-hard problems—level-3 expertise (Section 19.2)—requires a trained eye. This chapter provides this training, beginning with a single NP-hard problem (3-SAT) and concluding, after eighteen reductions, with a list of nineteen NP-hard problems, including all those studied earlier in this book. You can use this list as a starting point for NP-hardness proofs, and these reductions as templates for your own.

22.1 Reductions Revisited

What is NP-hardness, again? In Section 19.3.7, we provisionally defined an NP-hard problem as one for which a polynomial-time algorithm would refute the P $\neq$ NP conjecture, which in turn we informally described as the assertion that checking a solution to a problem (like a filled-out Sudoku puzzle) can be fundamentally easier than coming up with your own from scratch. (Chapter 23 is your source for 100% rigorous definitions.) Refuting the P $\neq$ NP conjecture would immediately solve thousands of problems—including almost all those studied in this book—that have resisted the efforts of countless brilliant minds over many decades. Thus, NP-hardness is strong evidence (if not an airtight proof) that a problem is intrinsically difficult and that the types of compromises described in Chapters 20 and 21 are required.

To apply the theory of NP-hardness, you don't actually have to understand any fancy mathematical definitions; this is one of the reasons why the theory has been adopted successfully far and wide, including broadly in engineering, the life sciences, and the social sciences.[1] The only prerequisite is the understanding of reductions that you already possess (Section 19.5.1):

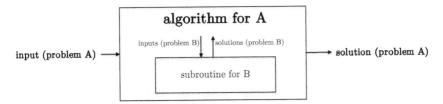

Formally, a problem A *reduces* to a problem B if A can be solved using a polynomial (in the input size) number of invocations of a subroutine that solves the problem B, along with a polynomial amount of additional work (outside of the subroutine calls). We've seen several examples of reductions that spread tractability from one problem (B) to another (A): If A reduces to B, a polynomial-time algorithm solving B automatically produces one for A (simply run the reduction, invoking the assumed subroutine for B as needed).

An NP-hardness proof turns this implication on its head, using a reduction for the nefarious purpose of spreading intractability from one problem to another (in the opposite direction of tractability):

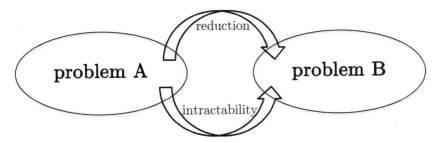

For if an NP-hard problem A reduces to B, any polynomial-time algorithm for B would automatically produce one for A, thereby refuting the P $\neq$ NP conjecture. That is, B must itself be NP-hard.

So, how do you prove that a problem is NP-hard? Just follow the simple two-step recipe.

---

[1]To see what I mean, check out how many results you get back from a search for "NP-hard" or "NP-complete" in your favorite academic database!

How to Prove a Problem Is NP-Hard

To prove that a problem B is NP-hard:

1. Choose an NP-hard problem A.

2. Prove that A reduces to B.

The rest of this chapter builds up your inventory of NP-hard problems (that is, choices for A in the first step) and hones your reduction skills (to be put to work in the second step).

A typical NP-hard problem B can be proved NP-hard using any number of choices for the known NP-hard problem A in the first step. The more closely A resembles B, the simpler the details of the second step. For example, the reduction in Section 19.5.4 from the directed Hamiltonian path problem to the cycle-free shortest path problem is relatively straightforward due to the similarities between the two problems.

22.2 3-SAT and the Cook-Levin Theorem

Every application of our two-step recipe identifies one new NP-hard problem using one old one. Apply it thousands of times and you'll have a catalog of thousands of NP-hard problems. But how does the process get started in the first place? With one of the most famous and important results in all of computer science: the *Cook-Levin theorem*, which proves from scratch that the seemingly innocuous 3-SAT problem is NP-hard.[2]

---

[2]Proved independently around 1971 by Stephen A. Cook and Leonid Levin on opposite sides of the Iron Curtain (Toronto and Moscow, respectively), although it took awhile for Levin's work to be widely appreciated in the West. Both hinted at the possibility that many more fundamental problems would be NP-hard. This prophecy was fulfilled by Richard M. Karp, who in 1972 demonstrated the power and reach of NP-hardness by using the two-step recipe to prove that an unexpectedly diverse array of notorious problems are NP-hard. Karp's work made clear that NP-hardness was the fundamental obstacle impeding algorithmic progress in many different directions. His original list of twenty-one NP-hard problems includes most of those studied in this chapter.

Cook and Karp were awarded the ACM Turing Award—the equivalent of the Nobel Prize in computer science—in 1982 and 1985, respectively. Levin was awarded the Knuth Prize—a lifetime achievement award in theoretical computer science—in 2012.

Theorem 22.1 (Cook-Levin Theorem) *The 3-SAT problem is NP-hard.*

The 3-SAT problem (introduced in Problem 21.13) is the special case of the SAT problem (Section 21.5) in which every constraint is a disjunction of at most three literals.[3,4]

Problem: 3-SAT

Input: A list of Boolean decision variables $x_1, x_2, \ldots, x_n$; and a list of constraints, each a disjunction of at most three literals.

Output: A truth assignment to $x_1, x_2, \ldots, x_n$ that satisfies every constraint, or a correct declaration that no such truth assignment exists.

For example, there's no way to satisfy all eight of the constraints

$$x_1 \vee x_2 \vee x_3 \quad x_1 \vee \neg x_2 \vee x_3 \quad \neg x_1 \vee \neg x_2 \vee x_3 \quad x_1 \vee \neg x_2 \vee \neg x_3$$
$$\neg x_1 \vee x_2 \vee x_3 \quad x_1 \vee x_2 \vee \neg x_3 \quad \neg x_1 \vee x_2 \vee \neg x_3 \quad \neg x_1 \vee \neg x_2 \vee \neg x_3,$$

as each of them forbids one of the eight possible truth assignments. If some constraint is removed, there is then one truth assignment left over that satisfies the other seven constraints. 3-SAT instances with and without a satisfying truth assignment are called *satisfiable* and *unsatisfiable*, respectively.

The 3-SAT problem occupies a central position in the theory of NP-hardness, both for historical reasons and because of the problem's equipoise between expressiveness and simplicity. To this day, the 3-SAT problem remains the most common choice for the known NP-hard problem in NP-hardness proofs (that is, for the problem A in the two-step recipe).

---

[3]Why three? Because this is the smallest value of k for which the k-SAT problem is NP-hard (see Problem 21.12).

[4]There is no contradiction between the Cook-Levin theorem and the remarkable successes of SAT solvers (Section 21.5). SAT solvers are only semi-reliable, solving some but not all SAT instances in a reasonable amount of time. They do not show that SAT is a polynomial-time solvable problem, so the P $\neq$ NP conjecture lives on!

In this chapter, we'll take the Cook-Levin theorem on faith. Standing on the shoulders of these giants, we'll assume only that a single problem (3-SAT) is NP-hard and then generate, via reductions, eighteen additional NP-hard problems. Section 23.3.5 outlines the high-level idea behind the proof of the Cook-Levin theorem and provides pointers for learning more.[5]

22.3   The Big Picture

We have a lot of problems and a lot of reductions between them to keep track of. Let's get organized.

22.3.1   A Rookie Mistake Revisited

As algorithm designers, we're accustomed to honorable reductions that spread tractability from one problem to another. Reductions nefariously spread intractability in the *opposite* direction, and for this reason, there's an overwhelming temptation to design reductions in the wrong direction (also known as the fifth rookie mistake from Section 19.6).

Quiz 22.1

Section 21.4 proves that the knapsack problem reduces to the mixed integer programming (MIP) problem. What does this imply? (Choose all that apply.)

a) If the MIP problem is NP-hard, so is the knapsack problem.

b) If the knapsack problem is NP-hard, so is the MIP problem.

c) A semi-reliable MIP solver can be translated to a semi-reliable algorithm for the knapsack problem.

---

[5]The proof is worth seeing at least once in your life, but almost nobody remembers the gory details. Most computer scientists are content to be educated clients of the Cook-Levin theorem, using it (and other NP-hard problems) like we do in this chapter, as a tool to prove problems NP-hard.

> d) A semi-reliable algorithm for the knapsack problem can be translated to a semi-reliable MIP solver.
>
> (See Section 22.3.4 for the solution and discussion.)

22.3.2   Eighteen Reductions

Figure 22.1 summarizes eighteen reductions, which (assuming the Cook-Levin theorem) imply that all nineteen problems in the figure are NP-hard.[6]

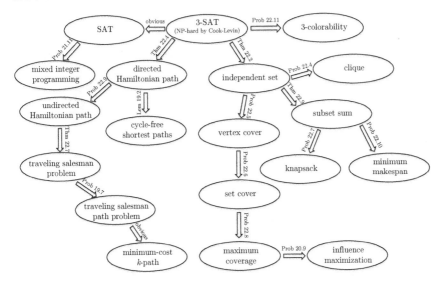

Figure 22.1: Eighteen reductions and nineteen NP-hard problems. An arrow from a problem A to a problem B indicates that A reduces to B. Computational intractability spreads in the same direction as the reductions, from the 3-SAT problem (which is NP-hard by the Cook-Levin theorem) to the other eighteen problems.

Six of these reductions are either immediate or have been smuggled into previous chapters of the book.

---

[6]Something stronger is true and will be explained in Chapter 23: The "search versions" of almost all of these problems are "NP-complete," and as a consequence any one of them can encode any other. The distinction between "NP-hard" and "NP-complete" is not of first-order importance to the algorithm designer: Either way, the problem is not polynomial-time solvable (assuming the P $\neq$ NP conjecture).

Reductions We've Already Seen

1. The 3-SAT problem is a special case of the general SAT problem (page 136) and thus trivially reduces to it.

2. The traveling salesman path problem (Problem 19.7 on page 40) trivially reduces to the minimum-cost k-path problem (page 114), as it is the special case in which the path length k equals the number of vertices.

3. Lemma 19.2 in Section 19.5.4 proves that the directed Hamiltonian path problem (page 30) reduces to the cycle-free shortest path problem (page 29).

4. Problem 19.7 proves that the traveling salesman problem (TSP; page 3) reduces to the traveling salesman path problem.

5. Problem 20.9 proves that the maximum coverage problem (page 56) reduces to the influence maximization problem (page 69).

6. Problem 21.11 proves that the SAT problem reduces to the mixed integer programming problem (page 132).

The end-of-chapter problems cover eight of the easier reductions.

Some Easier Reductions

7. Problem 22.4: The independent set problem (page 158) reduces to the clique problem (page 180).

8. Problem 22.5: The independent set problem reduces to the vertex cover problem (Problem 20.4 on page 97).

9. Problem 22.6: The vertex cover problem reduces to the set cover problem (Problem 20.2 on page 95).

10. Problem 22.7: The subset sum problem (page 172)

> reduces to the knapsack problem (page 19).
>
> 11. Problem 22.8: The set cover problem reduces to the maximum coverage problem.
>
> 12. Problem 22.9: The directed Hamiltonian path problem reduces to the undirected Hamiltonian path problem (page 169).
>
> 13. Problem 22.10: The subset sum problem reduces to the makespan minimization problem (page 42).
>
> 14. Problem 22.11: The 3-SAT problem reduces to the problem of checking whether a graph is 3-colorable (page 135).[7]

We're left holding a to-do list comprising four of the more difficult reductions:

> **Some Harder Reductions**
>
> 15. The 3-SAT problem reduces to the independent set problem (Section 22.5).
>
> 16. The 3-SAT problem reduces to the directed Hamiltonian path problem (Section 22.6).
>
> 17. The undirected Hamiltonian path problem reduces to the TSP (Section 22.7). (This one's not that difficult.)
>
> 18. The independent set problem reduces to the subset sum problem (Section 22.8).

22.3.3 Why Slog Through NP-Hardness Proofs?

I'll be honest: NP-hardness proofs can be painfully messy and problem-specific, and almost no one remembers their details. Why torture you with them over the next five sections? Because there are several good reasons to slog through a few:

---

[7] This reverses the direction of the reduction in Section 21.5.3; the intent is to spread (worst-case) intractability rather than (semi-reliable) tractability. Also, full disclosure: This reduction is somewhat harder than those in Problems 22.4–22.10.

Goals of Sections 22.4–22.8

1. Fulfill previously made promises that all the problems studied in this book are NP-hard and hence require the compromises described in Chapters 20 and 21.

2. Provide you with a long list of known NP-hard problems for use in your own reductions (in the first step of the two-step recipe).[8]

3. Empower you with the belief that, should the need arise, you could devise the reduction required to prove that a problem arising in your own work is NP-hard.

22.3.4   Solution to Quiz 22.1

Correct answers: (b),(c). A reduction from a problem A to a problem B spreads tractability from B to A and intractability in the opposite direction, from A to B (as we saw way back in Figures 19.2 and 19.3). Taking A and B as the knapsack and MIP problems, respectively, the reduction in Section 21.4 transfers tractability from the MIP problem to the knapsack problem (hence (c) is correct) and intractability in the reverse direction (hence (b) is correct).

22.4 A Template for Reductions

Typical reductions in NP-hardness proofs follow a common template. In general, a reduction from a problem A to a problem B can be sophisticated, invoking an assumed subroutine for B any polynomial number of times and processing its responses in polynomial time in arbitrarily clever ways (Section 22.1). At the other extreme, what would a simplest-imaginable reduction look like?

If we believe that the problem A is NP-hard (and that the P $\neq$ NP conjecture is true), every reduction from A to B must use the assumed subroutine for B at least once; otherwise, the reduction

---

[8]For a *really* long list (with more than 300 NP-hard problems), check out the classic book *Computers and Intractability: A Guide to the Theory of NP-Completeness*, by Michael R. Garey and David S. Johnson (Freeman, 1979). Few computer science books from 1979 remain as useful as this one!

would constitute a stand-alone polynomial-time algorithm for A. And for typechecking purposes, the instance of problem A provided to the reduction may require preprocessing before it makes sense to feed into the subroutine for B—if, for example, the input is a graph and the subroutine is expecting only a list of integers. Similarly, the response spit out by the subroutine may require postprocessing before it makes sense as the output of the reduction:

algorithm for problem A

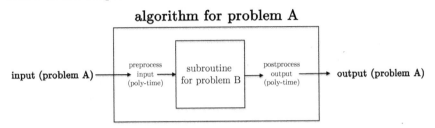

For our NP-hardness proofs, we'll be able to get away with these simplest-imaginable reductions:

Simplest-Imaginable Reduction from A to B

1. *Preprocessor:*   Given an instance of problem A, transform it in polynomial time to an instance of problem B.

2. *Subroutine:* Invoke the assumed subroutine for B.

3. *Postprocessor:*   Transform the subroutine's output in polynomial time to a correct output for the given instance of A.

The preprocessor and postprocessor are generally designed in tandem, with the former's transformation guided explicitly by the latter's needs. In all our examples, it will be obvious that the preprocessor and postprocessor run in polynomial (if not linear) time.

The reduction in Lemma 19.2, from the directed Hamiltonian path problem to the cycle-free shortest path problem, is archetypal. That reduction uses a preprocessor that converts an instance of the former problem into one of the latter by reusing the same graph and assigning each edge a length of -1, and a postprocessor that immediately deduces the correct output from the output of the assumed cycle-free shortest path subroutine. The reductions in the next four sections are more complex variations of the same idea.

22.5   Independent Set Is NP-Hard

The first NP-hardness proof of this chapter is for the *independent set* problem, in which the input is an undirected graph $G = (V, E)$ and the goal is to compute an independent set (that is, a set of mutually non-adjacent vertices) with the maximum-possible size.[9] For example, if G is a cycle graph with n vertices, the maximum size of an independent set is $n/2$ (if n is even) or $(n - 1)/2$ (if n is odd). If edges represent conflicts between people or tasks, independent sets correspond to the conflict-free subsets.

We currently have only one NP-hard problem at our disposal, the 3-SAT problem. Thus, if we're to prove that the independent set problem is NP-hard using the two-step recipe, our hand is forced: it can only be via a reduction from the 3-SAT problem to the independent set problem. These two problems seem to have nothing to do with each other: one is about logic and the other concerns graphs. Nevertheless, the main result of this section is:

Theorem 22.2 (Reduction from 3-SAT to Independent Set)
The 3-SAT problem reduces to the independent set problem.

Using the two-step recipe, because the 3-SAT problem is NP-hard (Theorem 22.1), so is the independent set problem.

Corollary 22.3 (NP-Hardness of Independent Set) *The independent set problem is NP-hard.*

22.5.1   The Main Idea

The reduction from the directed Hamiltonian path problem to the cycle-free shortest path problem in Lemma 19.2 exploited the strong similarities between the two problems, both of which are about finding paths in directed graphs. The 3-SAT and independent set problems, on the other hand, appear to be totally unrelated. If we're shooting for a simplest-imaginable reduction (Section 22.4), what's our plan for the preprocessor and postprocessor? The postprocessor must somehow extract a satisfying truth assignment for a 3-SAT instance

---

[9]This problem is the special case of the weighted independent set problem (page 19) in which every vertex weight is 1. Because this special case is NP-hard (as we'll see), so is the more general problem.

(or conclude that none exist) from a maximum-size independent set of a graph fabricated by the preprocessor. Next, we illustrate the main ideas through an example; Section 22.5.2 provides the formal description of the reduction and its proof of correctness.

To explain the reduction's preprocessor, think of a disjunction of k literals as someone's list of requests for their k favorite variable assignments. For example, the constraint $\neg x_1 \vee x_2 \vee x_3$ pleads: "could you set x_1 to false?"; "or what about x_2 to true?"; "or at least x_3 to true?" Meet at least one of their demands and they walk away happy, the constraint satisfied.

The key idea in the preprocessor is to encode an instance of the 3-SAT problem as a graph in which each vertex represents one assignment request by one constraint.[10] For example, the constraints

$$\underbrace{x_1 \vee x_2 \vee x_3}_{C_1} \qquad \underbrace{\neg x_1 \vee x_2 \vee x_3}_{C_2} \qquad \underbrace{\neg x_1 \vee \neg x_2 \vee \neg x_3}_{C_3}$$

would be represented by three groups of three vertices each:

$$\begin{array}{ccc}
\left(\begin{array}{c}(C_1)\\x_1=T\end{array}\right) \left(\begin{array}{c}(C_1)\\x_2=T\end{array}\right) \left(\begin{array}{c}(C_1)\\x_3=T\end{array}\right) &
\left(\begin{array}{c}(C_2)\\x_1=F\end{array}\right) \left(\begin{array}{c}(C_2)\\x_2=T\end{array}\right) \left(\begin{array}{c}(C_2)\\x_3=T\end{array}\right) &
\left(\begin{array}{c}(C_3)\\x_1=F\end{array}\right) \left(\begin{array}{c}(C_3)\\x_2=F\end{array}\right) \left(\begin{array}{c}(C_3)\\x_3=F\end{array}\right)
\end{array}$$

The fourth vertex, for instance, encodes the second constraint's plea to set the variable x_1 to false (corresponding to its literal $\neg x_1$).

Looking toward the postprocessor, do subsets of these vertices encode truth assignments? Not always. The issue is that some of the requests are inconsistent and ask for opposite assignments to the same variable (like the first and fourth vertices above). But remember, the whole point of the independent set problem is to represent conflicts! The preprocessor should therefore add an edge between each vertex pair corresponding to inconsistent assignments; because every independent set must choose at most one endpoint per edge, all conflicts are then avoided. Applying this idea to our running example (with the dashed vertices indicating one particular independent set):

---

[10]Your first thought might have been to turn a 3-SAT instance with n variables into a graph with n vertices, with the 2^n vertex subsets corresponding to the 2^n possible truth assignments. Alas, this approach doesn't pan out, motivating the more clever construction used here.

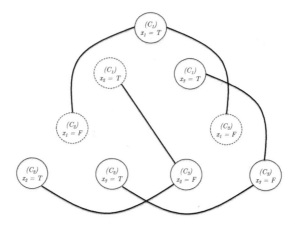

A postprocessor can now extract a satisfying truth assignment from any independent set S that contains at least one vertex in each group, simply by making all the corresponding variable assignment requests. (A variable with no requests in either direction can be safely assigned either true or false.) Because all the vertices of S are non-adjacent, none of the requests conflict, and the result is a well-defined truth assignment. Because S includes at least one vertex per group (one fulfilled request per constraint), this truth assignment satisfies all the constraints. For example, the three dashed vertices above would translate to one of the two satisfying assignments {false, true, true} or {false, true, false}.

Finally, the reduction must also recognize unsatisfiable 3-SAT instances. As we'll see in the next section, the preprocessor can make unsatisfiability obvious to the postprocessor by adding an edge between each pair of vertices that belong to the same group:

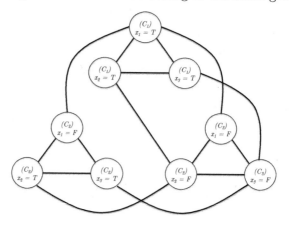

22.5.2   Proof of Theorem 22.2

The proof of Theorem 22.2 simply scales up the example and reasoning in Section 22.5.1 to general 3-SAT instances.

Description of the Reduction

Preprocessor.   Given an arbitrary 3-SAT instance, with n variables and m constraints with at most three literals each, the preprocessor constructs a corresponding graph $G = (V, E)$. It defines $V = V_1 \cup V_2 \cup \cdots \cup V_m$, where V_j is a group with one vertex per literal of the jth constraint. It defines $E = E_1 \cup E_2$, where E_1 contains one edge per pair of vertices that reside in the same group and E_2 contains one edge per pair of conflicting vertices (corresponding to requests for opposite assignments to the same variable).

Postprocessor.   If the assumed subroutine returns an independent set of the graph G constructed by the preprocessor with at least m vertices, the postprocessor returns an arbitrary truth assignment consistent with the corresponding variable assignment requests. Otherwise, the postprocessor returns "no solution."

Proof of Correctness

The crux of the correctness proof is showing that the preprocessor translates satisfiable and unsatisfiable 3-SAT instances into graphs in which the maximum size of an independent set is equal to and less than m, respectively:

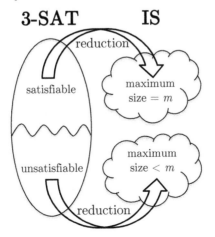

Case 1: Unsatisfiable instances. Suppose, for the purposes of a proof by contradiction, that the reduction fails to return "no solution" for some unsatisfiable 3-SAT instance. This means that the assumed subroutine returns an independent set S of the graph $G = (V, E)$ constructed by the preprocessor that includes at least m vertices. The edges of E_1 preclude more than one vertex from a group, so S must possess exactly m vertices, with one per group. Because of the edges in E_2, at least one truth assignment is consistent with all the assignment requests corresponding to the vertices of S. Because S includes one vertex from each group, the truth assignment extracted from S by the postprocessor would satisfy every constraint. This contradicts our initial assumption that the given 3-SAT instance is unsatisfiable.

Case 2: Satisfiable instances. Suppose the given 3-SAT instance has a satisfying truth assignment. Pick one fulfilled variable assignment request from each constraint—because the truth assignment satisfies every constraint, there must be one to pick—and let X denote the corresponding subset of m vertices. The set X is an independent set of G: It does not contain both endpoints of any edge of E_1 (as it contains only one vertex per group), nor of any edge of E_2 (as it is derived from a consistent truth assignment). With at least one size-m independent set of G out there to find, the assumed subroutine must return one—possibly X, or possibly some other size-m independent set (which in any case must have one vertex per group). As in case 1, the postprocessor then extracts from this independent set a satisfying assignment, which it returns as the reduction's (correct) output. $\mathcal{QED}$

Lest such correctness proofs strike you as overly pedantic, let's conclude this section with an example of a reduction gone awry.

Quiz 22.2

Where does the proof of Theorem 22.2 break down if the intragroup edges E_1 are omitted from the graph G? (Choose all that apply.)

   a) An independent set of G no longer translates to a

> well-defined truth assignment.
>
> b) A satisfiable 3-SAT instance need not translate to a graph in which the maximum size of an independent set is at least m, the number of constraints.
>
> c) An unsatisfiable 3-SAT instance need not translate to a graph in which the maximum size of an independent set is less than m.
>
> d) Actually, the proof still works.
>
> (See below for the solution and discussion.)

Correct answer: (c). With an unsatisfiable 3-SAT instance, and even without the intragroup edges E_1, no independent set of G includes one vertex from each of the m groups (as the postprocessor could translate any such independent set into a satisfying assignment). However, as independent sets of G are now free to recruit multiple vertices from a group, one may well include m vertices (or more).

*22.6   Directed Hamiltonian Path Is NP-Hard

With one reduction from the 3-SAT problem to a graph problem under our belt, why not another? In the *directed Hamiltonian path (DHP)* problem (page 30), the input is a directed graph $G = (V, E)$, a starting vertex $s \in V$, and an ending vertex $t \in V$. The goal is to return an s-t path visiting every vertex of G exactly once (called an s-t *Hamiltonian path*), or correctly declare that no such path exists.[11] In contrast to most of the nineteen problems studied in this chapter, our interest in this problem stems less from its direct applications and more from its utility in proving that other important problems (like the TSP) are NP-hard.

The main result of this section is:

Theorem 22.4 (Reduction from 3-SAT to DHP) *The 3-SAT problem reduces to the directed Hamiltonian path problem.*

---

[11]The problem statement on page 30 is slightly different, requiring only a "yes"/"no" answer rather than a path. Problem 22.3 asks you to show that the two versions of the problem are equivalent, with each reducing to the other.

Combined with the Cook-Levin theorem (Theorem 22.1) and our two-step recipe, Theorem 22.4 fulfills a promise made back in Section 19.5.4:

Corollary 22.5 (NP-Hardness of DHP) *The directed Hamiltonian path problem is NP-hard.*

22.6.1 Encoding Variables and Truth Assignments

To get away with a simplest-imaginable reduction (Section 22.4), we need a plan for the preprocessor (responsible for fabricating a directed graph from a 3-SAT instance) and the postprocessor (responsible for extracting a satisfying truth assignment from an *s-t* Hamiltonian path of that graph).

The first idea is to construct a graph in which an *s-t* Hamiltonian path is forced to make a sequence of binary decisions, which can then be interpreted as a truth assignment by the postprocessor. For example, in the diamond graph

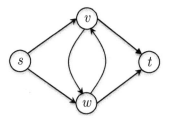

there are two *s-t* Hamiltonian paths: one that zig-zags downward $(s \to v \to w \to t)$ and one that zig-zags upward $(s \to w \to v \to t)$. Identifying down and up as "true" and "false," the *s-t* Hamiltonian paths encode the possible assignments to one Boolean variable.

What about more variables? The preprocessor will deploy one diamond graph per variable, chained together in a necklace. For example, the dashed *s-t* Hamiltonian path in the necklace graph

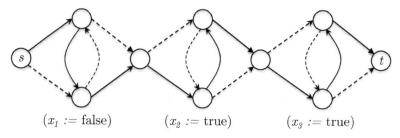

$(x_1 := \text{false})$ $(x_2 := \text{true})$ $(x_3 := \text{true})$

can be interpreted by the postprocessor as the truth assignment {false, true, true}; the rest of the s-t Hamiltonian paths similarly encode the other seven truth assignments.

22.6.2   Encoding Constraints

The preprocessor must next augment its graph to reflect the constraints of the given 3-SAT instance, so that only the satisfying truth assignments survive as s-t Hamiltonian paths. Here's an idea: Add one new vertex per constraint in such a way that visiting that vertex corresponds to satisfying the constraint. To see how this might work, consider the constraint $\neg x_1 \lor x_2 \lor x_3$ and the following graph (with the dashed edges indicating one particular s-t Hamiltonian path):

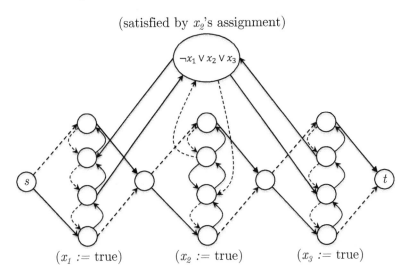

The edges between the necklace and the new constraint vertex allow visits to that vertex by s-t Hamiltonian paths only from diamonds that are traversed in a direction corresponding to a variable assignment that satisfies the constraint.[12]

For example, consider the dashed edges, an s-t Hamiltonian path. The path travels downward in each of the three diamonds, corresponding to the all-true truth assignment. Assigning x_1 to true does not

---

[12]Because every variable participates in this constraint, every diamond has an edge to and from the constraint vertex. If some variable were absent from the constraint, the corresponding diamond would have no such edges.

satisfy the constraint $\neg x_1 \lor x_2 \lor x_3$. Accordingly, there is no way to visit the new constraint vertex from the first diamond without skipping or visiting twice some vertex. Assigning x_2 to true does satisfy the constraint, which is why the dashed path can take a quick back-and-forth day trip to the constraint vertex from the second diamond before resuming its downward journey where it left off. Because x_3's assignment also satisfies the constraint, such a day trip is possible also from the third diamond. With the constraint vertex already visited, however, the s-t Hamiltonian path instead proceeds directly down the third diamond and over to t. (There's also a second s-t Hamiltonian path corresponding to the same truth assignment, traveling straight down in the second diamond and making the day trip from the third.)

To encode a second constraint, say $x_1 \lor \neg x_2 \lor \neg x_3$, the preprocessor can add another new vertex and wire it to the necklace in the same way (with the dashed edges indicating one particular s-t Hamiltonian path):

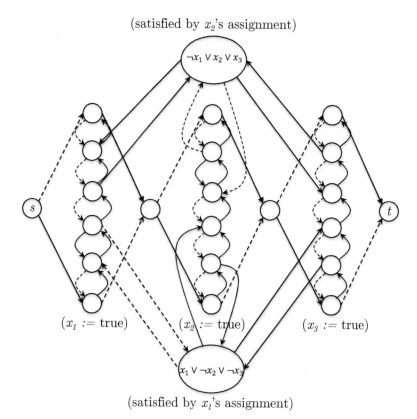

The two new vertices in each diamond provide room for any s-t Hamiltonian paths that might want to make back-and-forth day trips to both constraint vertices from the same diamond.[13] The dashed path is one of the two s-t Hamiltonian paths corresponding to the all-true truth assignment; the only opportunity to visit the new constraint vertex is from the first diamond. Of the other seven truth assignments, the five satisfying ones each correspond to one or more s-t Hamiltonian paths, while the other two do not.

22.6.3   Proof of Theorem 22.4

The proof of Theorem 22.4 scales up the example in Section 22.6.2 to general 3-SAT instances.

Description of the Reduction

Preprocessor.   Given a 3-SAT instance with n variables and m constraints, the preprocessor constructs a directed graph:

- Define a set V of $2mn + 3n + m + 1$ vertices: a starting vertex s; 3 external diamond vertices v_i, w_i, t_i for each variable x_i; $2m$ internal diamond vertices $a_{i,1}, a_{i,2}, \ldots, a_{i,2m}$ for each variable i; and m constraint vertices $c_1, c_2, \ldots, c_m$.

- Define a set E_1 of necklace edges by connecting s to v_1 and w_1; t_i to v_{i+1} and w_{i+1} for each $i = 1, 2, \ldots, n-1$; v_i and w_i to t_i, v_i to and from $a_{i,1}$, and w_i to and from $a_{i,2m}$ for each $i = 1, 2, \ldots, n$; and $a_{i,j}$ to and from $a_{i,j+1}$, for each $i = 1, 2, \ldots, n$ and $j = 1, 2, \ldots, 2m-1$.

- Define a set E_2 of constraint edges by connecting $a_{i,2j-1}$ to c_j and c_j to $a_{i,2j}$ whenever the jth constraint includes the literal x_i (that is, requests $x_i = $ true); and $a_{i,2j}$ to c_j and c_j to $a_{i,2j-1}$ whenever the jth constraint includes the literal $\neg x_i$ (requesting $x_i = $ false).

The preprocessor concludes with the graph $G = (V, E_1 \cup E_2)$; the starting and ending vertices of the constructed instance are defined as s and t_n, respectively.

---

[13]There are no such paths in this example, but there would be if we changed the second constraint to, say, $x_1 \lor \neg x_2 \lor x_3$.

Postprocessor. If the assumed subroutine computes an s-t_n Hamiltonian path P of the graph G constructed by the preprocessor, the postprocessor returns the truth assignment in which a variable x_i is set to true if P visits the vertex v_i before w_i and to false otherwise. If the assumed subroutine responds "no solution," the postprocessor also responds "no solution."

Proof of Correctness

The crux of the correctness proof is showing that the preprocessor translates satisfiable and unsatisfiable 3-SAT instances into graphs with and without an s-t_n Hamiltonian path, respectively:

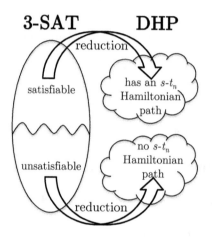

Case 1: Unsatisfiable instances. Suppose that the reduction fails to return "no solution" for some unsatisfiable 3-SAT instance. This means that the assumed subroutine returns an s-t_n Hamiltonian path P of the graph G constructed by the preprocessor. The Hamiltonian path P must resemble those in Section 22.6.2, traversing every diamond upward or downward and also visiting every constraint vertex. To visit a constraint vertex, the path must include a back-and-forth day trip interrupting some diamond traversal in a direction that corresponds to one of the constraint's variable assignment requests. (If the path fails to immediately return from the constraint vertex to the same diamond, it's out of options to visit the rest of the diamond later without visiting some vertex twice.) The truth assignment extracted from P by the postprocessor would therefore be

a satisfying assignment, contradicting the assumption that the given 3-SAT instance is unsatisfiable.

Case 2: Satisfiable instances. Suppose the given 3-SAT instance has a satisfying truth assignment. The graph G constructed by the preprocessor then has an s-t_n Hamiltonian path: traverse each diamond in the direction suggested by this assignment (downward for variables set to true, upward for the rest), taking a back-and-forth day trip to each constraint vertex at the earliest opportunity (from the diamond corresponding to the first variable whose assignment satisfies that constraint). With at least one s-t_n Hamiltonian path of G out there to find, the assumed subroutine must return one. As in case 1, the postprocessor then extracts from this path and returns a satisfying assignment. $\mathcal{QED}$

22.7　The TSP Is NP-Hard

We now return to a problem that we care about in its own right: the traveling salesman problem (TSP) from Section 19.1.2.

22.7.1　The Undirected Hamiltonian Path Problem

The plan is to piggyback on our hard work showing that the directed Hamiltonian path problem is NP-hard (Corollary 22.5), loosely following our reduction in Section 19.5.4 from that problem to the cycle-free shortest path problem. There is an immediate typechecking error, however, because the TSP concerns undirected rather than directed graphs. The *undirected* version of the Hamiltonian path problem seems more germane.

Problem: Undirected Hamiltonian Path (UHP)

Input: An undirected graph $G = (V, E)$, a starting vertex $s \in V$, and an ending vertex $t \in V$.

Output: An s-t path of G that visits every vertex exactly once (that is, an s-t Hamiltonian path), or a correct declaration that no such path exists.

Problem 22.9 asks you to show that the undirected and directed Hamiltonian path problems are equivalent, with each reducing to the other. Corollary 22.5 thus carries over to undirected graphs as well.

Corollary 22.6 (NP-Hardness of UHP) *The undirected Hamiltonian path problem is NP-hard.*

The main result in this section is then:

Theorem 22.7 (Reduction from UHP to TSP) *The undirected Hamiltonian path problem reduces to the traveling salesman problem.*

Combining this reduction with Corollary 22.6 shows that the TSP is indeed an NP-hard problem.

Corollary 22.8 (NP-Hardness of the TSP) *The traveling salesman problem is NP-hard.*

22.7.2   Proof of Theorem 22.7

We should breathe a sigh of relief that Theorem 22.7, unlike Theorems 22.2 and 22.4, relates two intuitively similar problems, both of which more or less concern long paths in undirected graphs. Given an undirected Hamiltonian path instance, how can we convert it to an instance of the TSP, so that a Hamiltonian path (or a correct declaration that none exist) can be easily extracted from a minimum-cost traveling salesman tour? The main idea is to simulate missing edges by costly edges.

Description of the Reduction

Preprocessor.   Given an undirected graph $G = (V, E)$, a starting vertex s, and an ending vertex t, the preprocessor first bridges the gap between paths that visit all vertices and cycles that visit all vertices by augmenting G with an additional vertex v_0 and edges connecting v_0 to s and t. It then assigns a cost of 0 to all edges in this augmented graph. To complete the construction of the TSP instance, the preprocessor adds in all the missing edges (to form the complete graph G' with vertex set $V \cup \{v_0\}$) and assigns each of these a cost of 1.

For example, the preprocessor translates the following graph with no *s*-*t* Hamiltonian path to a TSP instance with no zero-cost tour:

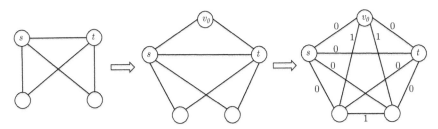

Postprocessor. If the assumed subroutine computes a zero-cost traveling salesman tour T of the graph G' constructed by the preprocessor, the postprocessor removes v_0 and its two incident edges from T and returns the resulting path. Otherwise, as in the example above, the postprocessor reports "no solution."

Proof of Correctness

To argue correctness, we'll justify this picture:

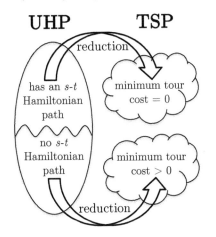

Case 1: Non-Hamiltonian instances. Suppose that the reduction fails to return "no solution" for some undirected Hamiltonian path instance G that has no *s*-*t* Hamiltonian path. This means that the assumed subroutine returns a zero-cost tour T of the graph G' constructed by the preprocessor—a tour that avoids all the cost-1 edges in G'. Because only the edges of G and the edges (v_0, s) and (v_0, t) have cost zero in G', the two edges of T incident to v_0 must be (v_0, s)

and (v_0, t), and the rest of T must be a cycle-free s-t path that visits all vertices of V while using only edges in G. Thus $T - \{(v_0, s), (v_0, t)\}$ is an s-t Hamiltonian path of G, contradicting our original assumption.

Case 2: Hamiltonian instances. Suppose the given undirected Hamiltonian path instance has an s-t Hamiltonian path P. The TSP instance G' constructed by the preprocessor then has a zero-cost tour in $P \cup \{(v_0, s), (v_0, t)\}$. With at least one zero-cost tour out there to find, the assumed subroutine must return one. As in case 1, the postprocessor then extracts from this tour and returns an s-t Hamiltonian path of G. $\mathcal{QED}$

22.8   Subset Sum Is NP-Hard

Last in our parade of NP-hardness proofs is one for the *subset sum* problem; as a consequence, both the knapsack and makespan minimization problems are also NP-hard (see Problems 22.7 and 22.10).

Problem: Subset Sum

Input: Positive integers $a_1, a_2, \ldots, a_n$, and a positive integer t.

Output: A subset of the a_i's with sum equal to t. (Or, correctly declare that no such subset exists.)

For example, if the a_i's are all the powers of ten from 1 to 10^{100}, there is a subset with a target sum t if and only if t (written base-10) has at most 101 digits, with each digit a 0 or a 1.

All the subset sum problem worries about is a bunch of numbers; it would seem to have nothing to do with problems that concern more complex objects like graphs. Nonetheless, the main result of this section is:

Theorem 22.9 (Reduction from IS to Subset Sum) *The independent set problem reduces to the subset sum problem.*

This result, in conjunction with Corollary 22.3, shows that:

Corollary 22.10 (NP-Hardness of Subset Sum) *The   subset sum problem is NP-hard.*

22.8.1   The Basic Approach

For now, let's focus on the problem of checking whether a given graph has an independent set of a given target size k, as opposed to computing a maximum-size independent set. (Any solution to the former problem extends easily to the latter: Use linear or binary search to identify the largest value of k for which the graph has a size-k independent set.)

The reduction's preprocessor must somehow metamorphose a graph and a target size into what our assumed subroutine for the subset sum problem is expecting: a bunch of positive integers.[14] The simplest-imaginable approach would define one number per vertex (along with a target t) so that independent sets of a target size k correspond to subsets of numbers that sum to t:

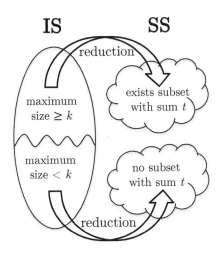

---

[14]As a special case of the knapsack problem (see Problem 22.7), the subset sum problem can be solved by dynamic programming in *pseudopolynomial* time, meaning in time polynomial in the input size and the magnitudes of the input numbers (see page 19 and Problem 20.11(a)). We should therefore expect a preprocessor to construct a subset sum instance with exponentially large numbers—an instance for which our dynamic programming algorithms offer no improvement over exhaustive search.

Problems that are both NP-hard and pseudopolynomial-time solvable are called *weakly NP-hard*, while *strongly NP-hard* problems remain NP-hard in instances with all input numbers bounded by a polynomial function of the input size. (An NP-hard problem with no numbers in the input, such as the 3-SAT problem, is automatically strongly NP-hard.) Of the nineteen problems in Figure 22.1, all but the subset sum and knapsack problems are strongly NP-hard.

22.8.2   Example: The Four-Cycle

The key idea is to use each of the lower-order digits of a number to encode whether an edge is incident to the corresponding vertex. For instance, the preprocessor could encode the vertices of the four-cycle

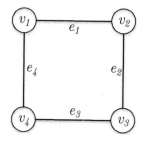

with the following five-digit numbers (written in base 10):

v_1	v_2	v_3	v_4
11,001	11,100	10,110	10,011

For example, the trailing four digits of v_2's encoding indicate that it is adjacent to e_1 and e_2 but not e_3 or e_4.

This idea shows promise. The two size-2 independent sets of the four-cycle, $\{v_1, v_3\}$ and $\{v_2, v_4\}$, correspond to two pairs of numbers with the same sum: $11{,}001 + 10{,}110 = 11{,}100 + 10{,}011 = 21{,}111$. All other subsets have different sums; for example, the sum corresponding to the non-independent set $\{v_3, v_4\}$ is $10{,}110 + 10{,}011 = 20{,}121$. A postprocessor could therefore translate any subset of numbers with sum 21,111 into a size-2 independent set of the four-cycle.

22.8.3   Example: The Five-Cycle

Suppose, however, that we try the same maneuver with the five-cycle:

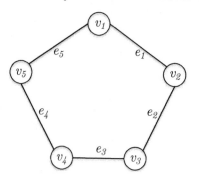

with each vertex (and its incident edges) encoded using a six-digit number. Different size-2 independent sets now correspond to pairs of numbers with different sums—211,101 and 211,110 for $\{v_1, v_3\}$ and $\{v_2, v_4\}$, for example. In general, a lower-order digit of the sum will be 0 if it corresponds to an edge with neither endpoint in the independent set, and 1 otherwise.

To correct lower-order digits that would otherwise be 0, the prepro-cessor can define one additional number per edge. For the five-cycle, the final list of numbers is:

v_1	v_2	v_3	v_4	v_5
110,001	111,000	101,100	100,110	100,011
e_1	e_2	e_3	e_4	e_5
10,000	1,000	100	10	1

Now, the target sum $t = 211{,}111$ can be achieved by taking the numbers corresponding to a size-2 independent set (like $\{v_1, v_3\}$ or $\{v_2, v_4\}$) along with the numbers corresponding to edges with neither endpoint in the independent set (like e_4 or e_5, respectively). There is no other way to achieve this target sum (as you should check).

22.8.4   Proof of Theorem 22.9

The proof of Theorem 22.9 scales up the example in the preceding section to general independent set instances.

Description of the Reduction

Preprocessor. Given both an undirected graph $G = (V, E)$ with vertex set $V = \{v_1, v_2, \ldots, v_n\}$ and edge set $E = \{e_1, e_2, \ldots, e_m\}$ and a target size k, the preprocessor constructs $n + m + 1$ positive integers that define an instance of the subset sum problem:

- For each vertex v_i, define the number $a_i = 10^m + \sum_{e_j \in A_i} 10^{m-j}$, where A_i denotes the edges incident to v_i. (Written in base 10, the leading digit is 1 and the jth digit after that is 1 if e_j is incident to v_i and 0 otherwise.)

- For each edge e_j, define the number $b_j = 10^{m-j}$.

- Define the target sum $t = k \cdot 10^m + \sum_{j=1}^{m} 10^{m-j}$. (Written in base 10, the digits of k followed by m 1's.)

Postprocessor. If the assumed subroutine computes a subset of $\{a_1, a_2, \ldots, a_n, b_1, b_2, \ldots, b_m\}$ with sum t, the postprocessor returns the vertices v_i that correspond to the a_i's in the subset. (For example, if handed the subset $\{a_2, a_4, a_7, b_3, b_6\}$, the postprocessor returns the vertex subset $\{v_2, v_4, v_7\}$.) If the assumed subroutine responds "no solution," the postprocessor also responds "no solution."

Outer loop. The preprocessor and postprocessor are designed to check for an independent set of a target size k. To compute a maximum-size independent set of an input graph $G = (V, E)$, the reduction checks all possible values of $k = n, n - 1, n - 2, \ldots, 2, 1$:

1. Invoke the preprocessor to transform G and the current value of k into an instance of the subset sum problem.

2. Invoke the assumed subroutine for the subset sum problem.

3. Invoke the postprocessor on the subroutine's output. If it returns a size-k independent set S of G, halt and return S.

Overall, the reduction invokes the subset sum subroutine at most n times and performs at most a polynomial amount of additional work.

Proof of Correctness

The reduction is correct provided that every iteration of its outer loop correctly determines, for the current value of k, whether the input graph has a size-k independent set.

Case 1: No size-k independent set. Suppose that an iteration of the reduction's outer loop fails to return "no solution" for some graph $G = (V, E)$ that has no size-k independent set. This means that the assumed subroutine returns a subset N of the numbers $\{a_1, a_2, \ldots, a_n, b_1, b_2, \ldots, b_m\}$ constructed by the preprocessor with the target sum t. Let $S \subseteq V$ denote the vertices that correspond to the a_i's in N. To obtain a contradiction, we next argue that S is a size-k independent set of G.

In general, for every subset of s of the a_i's and any number of the b_j's, the sum (written base-10) is the digits of s followed by m digits that each belong to $\{0, 1, 2, 3\}$. (Exactly three numbers can contribute to the jth of the m trailing digits: b_j, and the two a_i's that correspond to e_j's endpoints.) Because the leading digits of the target sum t match those of k, the subset N must contain k of the a_i's; thus, S has size k. Because the m trailing digits of t are all 1, the subset N cannot contain two a_i's that correspond to the endpoints of a common edge; thus, S is an independent set of G.

Case 2: At least one size-k independent set. Suppose the input graph $G = (V, E)$ has a size-k independent set S. The subset sum instance constructed by the preprocessor then has a subset with the target sum t: Choose the k a_i's corresponding to the vertices of S, along with the b_j's corresponding to the edges with neither endpoint in S. With at least one feasible solution out there to find, the assumed subroutine must return a subset N with sum t. As in case 1, the postprocessor then extracts from N and returns a size-k independent set of G. *QED*

The Upshot

☆ To prove that a problem B is NP-hard, follow the two-step recipe: (i) choose an NP-hard problem A; (ii) prove that A reduces to B.

☆ The 3-SAT problem is the special case of the satisfiability problem in which every constraint has at most three literals.

☆ The Cook-Levin theorem proves that the 3-SAT problem is NP-hard.

☆ Starting from the 3-SAT problem, thousands of applications of the two-step recipe have proved that thousands of problems are NP-hard.

☆ Reductions in NP-hardness proofs conform to a template: preprocess the input; invoke the assumed subroutine; postprocess the output.

☆ In the independent set problem, the input is an undirected graph and the goal is to compute a maximum-size subset of non-adjacent vertices.

☆ The 3-SAT problem reduces to the independent set problem, proving the latter NP-hard.

☆ An s-t Hamiltonian path of a graph G starts at the vertex s, ends at the vertex t, and visits every vertex of G exactly once.

☆ The 3-SAT problem reduces to the directed version of the Hamiltonian path problem, proving the latter NP-hard.

☆ The undirected version of the Hamiltonian path problem reduces to the traveling salesman problem, proving the latter NP-hard.

☆ In the subset sum problem, the goal is to compute a subset of a given set of positive integers with sum equal to a given target (or conclude that none exist).

☆ The independent set problem reduces to the subset sum problem, proving the latter NP-hard.

Test Your Understanding

Problem 22.1 *(S)* Assume that the P $\neq$ NP conjecture is true. Which of the following problems can be solved in polynomial time? (Choose all that apply.)

a) Given a connected undirected graph, compute a spanning tree with the smallest-possible number of leaves.

b) Given a connected undirected graph, compute a spanning tree with the minimum-possible maximum degree. (The degree of a vertex is the number of incident edges.)

c) Given a connected undirected graph with nonnegative edge lengths, a starting vertex s, and an ending vertex t, compute the minimum length of a cycle-free s-t path with exactly $n - 1$ edges (or $+\infty$, if no such path exists).

d) Given a connected undirected graph with nonnegative edge lengths, a starting vertex s, and an ending vertex t, compute the minimum length of a (not necessarily cycle-free) s-t path with exactly $n - 1$ edges (or $+\infty$, if no such path exists).

Problem 22.2 *(S)* Assume that the P $\neq$ NP conjecture is true. Which of the following problems can be solved in polynomial time? (Choose all that apply.)

a) Given a directed graph $G = (V, E)$ with nonnegative edge lengths, compute the longest length of a shortest path between any pair of vertices (that is, $\max_{v,w \in V} dist(v, w)$, where $dist(v, w)$ denotes the shortest-path distance from vertex v to vertex w).

b) Given a directed acyclic graph with real-valued edge lengths, compute the length of a longest path between any pair of vertices.

c) Given a directed graph $G = (V, E)$ with nonnegative edge lengths, compute the length of a longest cycle-free path between any pair of vertices (that is, $\max_{v,w \in V} maxlen(v, w)$, where $maxlen(v, w)$ denotes the length of a longest cycle-free path from v to w).

d) Given a directed graph with real-valued edge lengths, compute the length of a longest cycle-free path between any pair of vertices.

Problem 22.3 *(S)* Call the version of the directed Hamiltonian path problem on page 30, in which only a "yes"/"no" answer is required, the *decision version*. Call the version on page 163, in which an s-t Hamiltonian path itself is required (whenever one exists), the *search version*. Call the version of the TSP in Section 19.1.2 the *optimization version*, and define the *search version* of the TSP as: Given a complete graph, real-valued edge costs, and a target cost C, return a traveling salesman tour with a total cost of at most C (or correctly declare that none exist).

Which of the following are true? (Choose all that apply.)

a) The decision version of the directed Hamiltonian path problem reduces to the search version.

b) The search version of the directed Hamiltonian path problem reduces to the decision version.

c) The search version of the TSP reduces to the optimization version.

d) The optimization version of the TSP reduces to the search version.

Problem 22.4 *(H)* In the *clique* problem, the input is an undirected graph and the goal is to output a clique—a subset of mutually adjacent vertices—with the maximum-possible size. Prove that the independent set problem reduces to the clique problem, implying (by Corollary 22.3) that the latter is NP-hard.

Problem 22.5 *(H)* In the *vertex cover* problem, the input is an undirected graph $G = (V, E)$, and the goal is to identify a minimum-size subset $S \subseteq V$ of vertices that includes at least one endpoint of every edge in E. Prove that the independent set problem reduces to the vertex cover problem, implying (by Corollary 22.3) that the latter is NP-hard.

Problem 22.6 *(H)* In the *set cover* problem, the input comprises m subsets $A_1, A_2, \ldots, A_m$ of a ground set U, and the goal is to identify a minimum-size collection of subsets whose union equals U. Prove that the vertex cover problem reduces to the set cover problem, implying (by Problem 22.5) that the latter is NP-hard.

Problem 22.7 *(H)* Prove that the subset sum problem reduces to the knapsack problem (page 19), implying (by Corollary 22.10) that the latter is NP-hard.

Challenge Problems

Problem 22.8 *(S)* Prove that the set cover problem reduces to the maximum coverage problem (Section 20.2.1), implying (by Problem 22.6) that the latter is NP-hard.

Problem 22.9 *(H)* Prove that the undirected Hamiltonian path problem reduces to the directed Hamiltonian path problem and vice versa. (In particular, Corollary 22.6 follows from Corollary 22.5.)

Problem 22.10 *(H)*

(a) Prove that the subset sum problem remains NP-hard in the special case in which the target sum equals half the sum of the input numbers (that is, $t = \frac{1}{2}\sum_{i=1}^{n} a_i$).[15]

(b) Prove that this special case of the subset sum problem reduces to the makespan minimization problem with two machines, implying (by (a)) that the latter is NP-hard.[16]

Problem 22.11 *(H)* Prove that the 3-SAT problem reduces to the special case of the graph coloring problem (page 135) in which the number k of allowable colors is 3, implying (by the Cook-Levin theorem) that the latter is NP-hard.[17]

Problem 22.12 *(H)* Problem 20.12 introduced the *metric* special case of the TSP, in which the edge costs c of the input graph $G = (V, E)$ are nonnegative and satisfy the triangle inequality:

$$c_{vw} \leq \sum_{e \in P} c_e$$

for every pair $v, w \in V$ of vertices and v-w path P in G. Problem 20.12 also developed a polynomial-time heuristic algorithm that, given a metric TSP instance, is guaranteed to return a tour with total cost at most twice the minimum possible. Can we do better and solve the metric special case exactly, or at least extend the heuristic algorithm's approximate correctness guarantee to the general TSP?

(a) Prove that the metric special case of the TSP is NP-hard.

(b) Assume that the P $\neq$ NP conjecture is true. Prove that there is no polynomial-time algorithm that, for every TSP instance with nonnegative edge costs (and no other assumptions), returns a tour with total cost at most 10^{100} times the minimum possible.

---

[15]This special case of the subset sum problem is often called the *partition* problem.

[16]The two-machine special case of the makespan minimization problem can be solved in pseudopolynomial time by dynamic programming (as you should check) and is therefore only weakly NP-hard (see footnote 14). A more complicated reduction shows that the general version of the problem is strongly NP-hard.

[17]The graph coloring problem can be solved in linear time when $k = 2$ (see footnote 43 in Chapter 21).

Chapter 23

P, NP, and All That

Chapters 19–22 cover everything the pure algorithm designer needs to know about NP-hard problems—the algorithmic implications of NP-hardness, algorithmic tools for making headway on NP-hard problems, and how to spot NP-hard problems in the wild. We provisionally defined NP-hardness in terms of the P $\neq$ NP conjecture and informally described this conjecture in Section 19.3.5, without any formal mathematical definitions (which we didn't need at the time). This optional chapter fills in the missing foundations.[1]

Section 23.1 outlines our plan to amass evidence of a problem's intractability by reducing a large number of problems to it. Section 23.2 distinguishes three types of computational problems: decision, search, and optimization problems. Section 23.3 defines the complexity class $\mathcal{NP}$ as the set of all search problems with efficiently recognizable solutions, formally defines NP-hard problems, and revisits the Cook-Levin theorem. Section 23.4 formally defines the P $\neq$ NP conjecture and surveys its current status. Section 23.5 describes two important conjectures that are stronger than the P $\neq$ NP conjecture—the Exponential Time Hypothesis (ETH) and Strong Exponential Time Hypothesis (SETH)—and their algorithmic implications (for example, for the sequence alignment problem). Section 23.6 concludes with a discussion of Levin reductions and NP-complete problems—universal problems that simultaneously encode all problems with efficiently recognizable solutions.

---

[1] This chapter is an introduction to a beautiful and mathematically deep field called *computational complexity theory*, which studies the quantity of computing resources (like time, memory, or randomness) necessary to solve different computational tasks (as a function of the input size). We'll maintain a ruthless focus on the algorithmic implications of this theory, resulting in a slightly unconventional treatment. If you want to learn more about computational complexity theory, I recommend starting with Ryan O'Donnell's excellent (and freely available) video lectures (http://www.cs.cmu.edu/~odonnell/).

*23.1   Amassing Evidence of Intractability

In 1967, Jack Edmonds conjectured that the traveling salesman problem (TSP) cannot be solved by any polynomial-time algorithm, not even one with a running time of $O(n^{100})$ or $O(n^{10000})$ for inputs with n vertices (page 6). Absent a mathematical proof, how could we build a compelling case that this conjecture is true? The failed efforts to come up with such an algorithm by so many brilliant minds over the past seventy years constitute circumstantial evidence of intractability, but can we do better?

23.1.1   Building a Case with Reductions

The key idea is to show that a polynomial-time algorithm for the TSP would not merely solve one unsolved problem—it would solve *thousands* of them.

Amassing Evidence That the TSP Is Intractable

1. Choose a really big collection $\mathcal{C}$ of computational problems.

2. Prove that *every* problem in $\mathcal{C}$ reduces to the TSP.[2]

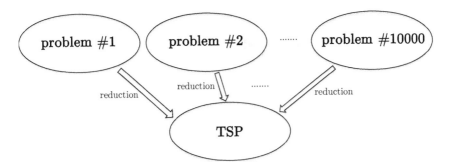

---

[2] As a reminder from Sections 19.5.1 and 22.1, a *reduction* from a problem A to a problem B is an algorithm that solves problem A while using at most a polynomial (in the input size) number of calls to a subroutine solving B and a polynomial amount of additional work. This type of reduction is sometimes called a *Cook reduction* (after Stephen Cook) or a *polynomial-time Turing reduction* (after Alan Turing), and is the most sensible one to focus on when studying algorithms. More restricted types of reductions are important for defining "NP-complete" problems; see Section 23.6.

A polynomial-time algorithm for the TSP would then automatically provide one for every problem in the set $\mathcal{C}$. Said another way, if even *one* problem in the set $\mathcal{C}$ cannot be solved by a polynomial-time algorithm, neither can the TSP. The bigger the set $\mathcal{C}$, the stronger the argument that the TSP is not a polynomial-time solvable problem.

23.1.2   Choosing the Set $\mathcal{C}$ for the TSP

To make the case against the TSP's polynomial-time solvability as compelling as possible, why not reach for the stars and take $\mathcal{C}$ to be the set of *all* the computational problems in the world? Because this is too ambitious. Hard as the TSP may be, there are computational problems out there that are much, much harder. At the extreme are *undecidable* problems—problems that cannot be solved by a computer in any finite amount of time (not even in exponential time, not even in doubly-exponential time, and so on). One famous example of an undecidable problem is the *halting problem*: Given a program (say, one thousand lines of Python), determine whether it goes into an infinite loop or eventually halts. The obvious approach is to run the program and see what it does. But if the program hasn't halted after a century, how do you know if it's in an infinite loop or if it will halt tomorrow? You might hope for some shortcut smarter than rote simulation of the code, but unfortunately none exist in general.[3]

The TSP no longer looks so bad—at least it can be solved in a finite (albeit exponential) amount of time via exhaustive search. No way can the halting problem reduce to the TSP, as such a reduction would translate the exponential-time algorithm for the TSP into one for the halting problem (which, per Turing, does not exist).

---

[3]In 1936, Alan M. Turing published his paper "On Computable Numbers, with an Application to the Entscheidungsproblem" (*Proceedings of the London Mathematical Society*, 1936). I and many other computer scientists view this paper as the birth of our discipline and, for this reason, believe that Turing's name should be as widely recognized as, say, that of Albert Einstein.

What made this paper so important? Two things. First, Turing introduced a formal model of what general-purpose computers can do, now called a *Turing machine*. (Mind you, this was ten years before anyone had actually built a general-purpose computer!) Second, defining what computers can do enabled Turing to study what they *can't* do and to prove that the halting problem is undecidable. Thus, from literally day 1, computer scientists have been acutely aware of computers' limitations and the necessity of compromise when tackling hard computational problems.

Going back to the drawing board, what is the largest set $\mathcal{C}$ of computational problems that might conceivably reduce to the TSP? Intuitively, the biggest set we could hope for is the set of *all problems solvable by an analogous exhaustive search algorithm*—then, among all such problems, the TSP would be the hardest one. Could there be a mathematical definition that captures this idea?

*23.2   Decision, Search, and Optimization

Before defining the set of "problems solvable by naive exhaustive-search"—the complexity class $\mathcal{NP}$—let's step back and categorize the different types of input-output formats that we've seen.

Three Types of Computational Problems

1. *Decision problem.* Output "yes" if there is a feasible solution and "no" otherwise.

2. *Search problem.* Output a feasible solution if one exists, and "no solution" otherwise.

3. *Optimization problem.* Output a feasible solution with the best-possible objective function value (or "no solution," if none exist).

Decision problems are the rarest of the three in applications, and we've seen only one example in this book: the original description of the directed Hamiltonian path problem (page 30), in which "feasible solutions" correspond to *s-t* Hamiltonian paths. Search and optimization problems are both common. Of the nineteen problems studied in Chapter 22 (see Figure 22.1), six are search problems: the 3-SAT, SAT, graph coloring, directed Hamiltonian path (the version on page 163), undirected Hamiltonian path, and subset sum problems. The other thirteen are optimization problems.[4] The definition of a "feasible solution"—such as a satisfying assignment, a Hamiltonian

---

[4]To shoehorn the cycle-free shortest path problem into this definition of an optimization problem, consider the variant in which two vertices are supplied as input and a shortest cycle-free path from the first to the second is required as output. The NP-hardness proof in Section 19.5.4 (Lemma 19.2) also applies to this version of the problem.

path, or a traveling salesman tour (perhaps with at most some target total cost)—is problem-specific. For the optimization problems, the objective function—such as minimizing the total cost or maximizing the total value—is also problem-specific.

Complexity classes usually stick with problems of only one type to avoid typechecking errors, and we'll restrict our definition of $\mathcal{NP}$ to search problems.[5] Don't worry about leaving optimization problems like the TSP out in the cold: Every optimization problem has a corresponding search version. The input to the search version also includes a target objective function value t; the goal is then to find a feasible solution with value at least t (for maximization problems) or at most t (for minimization problems), or correctly report that none exist. As we'll see, the search versions of almost all the optimization problems that we have studied reside in $\mathcal{NP}$.[6]

*23.3 $\mathcal{NP}$: Problems with Easily Recognized Solutions

We now arrive at the heart of the discussion. How can we define the set of "exhaustive-search-solvable" problems—the set of all problems that might plausibly reduce to the TSP? What are the minimal ingredients for solving a problem by naive exhaustive search?

23.3.1 Definition of the Complexity Class $\mathcal{NP}$

The big idea behind the complexity class $\mathcal{NP}$ is the *efficient recognition of purported solutions*. That is, if someone handed you an alleged feasible solution to a problem instance on a silver platter, you could quickly check whether it was indeed a feasible solution. For example,

---

[5]Most books define the complexity class $\mathcal{NP}$ in terms of decision problems; this is more convenient for developing complexity theory but further removed from natural algorithmic problems. The version of the class used here is sometimes called $\mathcal{FNP}$, where the "$\mathcal{F}$" stands for "functional." All the algorithmic implications of NP-hardness, including the truth or falsity of the P $\neq$ NP conjecture, remain the same no matter which definition is used.

[6]The search version of an optimization problem reduces immediately to the original version: Either an optimal solution meets a given target objective function value t, or no feasible solutions do. More interesting is the converse: A typical optimization problem reduces to its search version via binary search over the target t (see also Problem 22.3). Such an optimization problem is polynomial-time solvable if and only if its search version is polynomial-time solvable, and similarly is NP-hard if and only if its search version is NP-hard.

if someone hands you a filled-out Sudoko or KenKen puzzle, it's easy to check whether they followed all the rules. Or, if someone suggests a sequence of vertices in a graph, it's easy to check whether they constitute a traveling salesman tour and, if they do, whether the total cost of the tour is at most a given target t.[7]

The Complexity Class $\mathcal{NP}$

A search problem belongs to the complexity class $\mathcal{NP}$ if and only if:

1. For every instance, every candidate solution has description length (in bits, say) bounded above by a polynomial function of the input size.

2. For every instance and candidate solution, the alleged feasibility of the solution can be confirmed or denied in time polynomial in the input size.

23.3.2   Examples of Problems in $\mathcal{NP}$

The entrance requirements for membership in $\mathcal{NP}$ are so easily passed that almost all the search problems that you've seen qualify. For example, the search version of the TSP belongs to the class $\mathcal{NP}$: A tour of n vertices can be described using $O(n \log n)$ bits—roughly $\log_2 n$ bits to name each vertex—and, given a list of vertices, it's easy to check whether they constitute a tour with total cost at most a given target t. The 3-SAT problem (Section 22.2) also belongs to $\mathcal{NP}$: Describing a truth assignment to n Boolean variables takes n bits, and checking whether one satisfies each of the given constraints is straightforward. Similarly, it's easy to check whether a proposed path is Hamiltonian, whether a proposed job schedule has a given makespan, or whether a proposed subset of vertices is an independent set, vertex cover, or clique of a given size.

---

[7]$\mathcal{NP}$ can equivalently be defined as the search problems that are efficiently solvable in a fictitious computational model defined by "nondeterministic Turing machines." The acronym "NP" stands for "nondeterministic polynomial-time" (and not for "not polynomial"!) and refers to this alternative definition. In an algorithms context, you should always think of $\mathcal{NP}$ problems as those with efficiently recognizable solutions.

Quiz 23.1

Of the nineteen problems listed in Figure 22.1, for how many does the search version belong to $\mathcal{NP}$?

a) 16

b) 17

c) 18

d) 19

(See Section 23.3.6 for the solution and discussion.)

23.3.3 $\mathcal{NP}$ Problems Are Solvable by Exhaustive Search

We originally set out to define the set of problems that are solvable by naive exhaustive search—the problems with a shot at reducing to the TSP—but instead defined the class $\mathcal{NP}$ as the search problems with efficiently recognizable solutions. The connection? Every problem in $\mathcal{NP}$ can be solved in exponential time by using naive exhaustive search to check candidate solutions one by one:

Exhaustive Search for a Generic $\mathcal{NP}$ Problem

1. Enumerate candidate solutions, one by one:

   a) If the current candidate is feasible, return it.

2. Return "no solution."

For a problem in $\mathcal{NP}$, candidate solutions require $O(n^d)$ bits to describe, where n denotes the input size and d is a constant (independent of n). Thus, the number of possible candidates (and hence of loop iterations) is $2^{O(n^d)}$.[8] By the second defining property of an NP problem, each loop iteration can be carried out in polynomial time.

---

[8] Big-O notation in an exponent suppresses constant factors (and lower order terms) in the exponent. For example, a function $T(n)$ is $2^{O(\sqrt{n})}$ if there are constants $c, n_0 > 0$ such that $T(n) \leq 2^{c\sqrt{n}}$ for all $n \geq n_0$. (Whereas $T(n) = O(2^{\sqrt{n}})$ means that $T(n) \leq c \cdot 2^{\sqrt{n}}$ for all $n \geq n_0$.)

Exhaustive search therefore correctly solves the problem in time "only" exponential in the input size n.

23.3.4   NP-Hard Problems

The requirements for membership in the complexity class $\mathcal{NP}$ are extremely weak. All you need is the ability to recognize a correct solution—to know one when you see one. As a result, $\mathcal{NP}$ is an enormous class of search problems, capturing the overwhelming majority of those you're likely to encounter. So if *every* problem in $\mathcal{NP}$ reduces to a problem A—if A is at least as hard as every problem in $\mathcal{NP}$—a polynomial-time algorithm for A would lead directly to such algorithms for the entire gamut of $\mathcal{NP}$ problems. This constitutes strong evidence of intrinsic intractability and is exactly the formal definition of an *NP-hard* problem.

NP-Hard Problem (Formal Definition)

A computational problem is *NP-hard* if every problem in $\mathcal{NP}$ reduces to it.

Once we formally define the P $\neq$ NP conjecture in Section 23.4, we'll see that every problem that is NP-hard under this definition also satisfies the provisional definition (Section 19.3.7) used throughout Chapters 19–22. The minor switch in definition affects none of the lessons from those chapters. For example, the Cook-Levin theorem (Theorem 22.1) shows that the 3-SAT problem is NP-hard according to the formal definition (as we'll see in Section 23.3.5); reductions continue to spread NP-hardness (Problem 23.4); and as a consequence, the nineteen problems studied in Chapter 22 remain NP-hard under this new definition.[9]

---

[9]Look in other books and you'll often see a more demanding definition of NP-hardness that requires reductions of a very specific form, called "Levin reductions." (Section 23.6.1 defines such reductions and Section 23.6.2 uses them to define "NP-complete" problems.) Only search problems are eligible for NP-hardness under this more restrictive definition; instead of "the TSP is NP-hard," one must say "the search version of the TSP is NP-hard." The more liberal definition used here, with general (Cook) reductions, better accords with the algorithmic viewpoint of this book series.

23.3.5   The Cook-Levin Theorem Revisited

In Chapter 22 we were content to take the Cook-Levin theorem
(Theorem 22.1) on faith and couple it with our two-step recipe to
prove problems NP-hard. We're now equipped with all the definitions
necessary to understand precisely what that theorem says: Every
problem in $\mathcal{NP}$ reduces to the 3-SAT problem. How could this be
true? The 3-SAT problem seems so simple, the class $\mathcal{NP}$ so vast.

The details of the proof get messy, but here's the gist.[10] Fix an
arbitrary $\mathcal{NP}$ problem A; we must show that A reduces to the 3-SAT
problem. All we know about A is that it meets the two defining requirements of an $\mathcal{NP}$ problem: (i) feasible solutions to size-n instances
can be described using at most $c_1 n^{d_1}$ bits; and (ii) alleged feasible
solutions to size-n instances can be verified in at most $c_2 n^{d_2}$ time
(where c_1, c_2, d_1, d_2 are constants). Denote by $\mathtt{verify}_A$ the algorithm
in (ii) that checks the feasibility of a purported solution.

We'll be able to get away with a simplest-imaginable reduction
(Section 22.4). The main ingredient is a preprocessor that translates
instances of A with and without a feasible solution into satisfiable
and unsatisfiable 3-SAT instances, respectively:

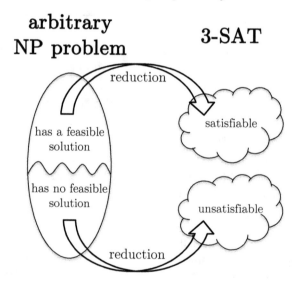

Preprocessor.   Given a size-n instance I_A of the problem A, the
preprocessor constructs a 3-SAT instance I_{3SAT}:

---

[10]For a complete proof, refer to any textbook on computational complexity.

- Define $c_1 n^{d_1}$ *solution* variables. The intent is for these variables to record bits that describe a candidate solution for I_A.

- Define $(c_2 n^{d_2})^2 = c_2^2 \cdot n^{2d_2}$ *state* variables. The intent is for these variables to encode the execution of $\mathtt{verify}_A$ on the candidate solution for I_A encoded by the solution variables.

- Define constraints to enforce the semantics of the state variables. A typical constraint asserts: "the jth bit of memory after step $i+1$ is consistent with the relevant memory contents after step i, the given instance I_A, the candidate solution to I_A encoded by the solution variables, and the code for the algorithm $\mathtt{verify}_A$."

- Define constraints that ensure that the computation encoded by the state variables concludes with an assertion of the feasibility of the candidate solution encoded by the solution variables.

Why $(c_2 n^{d_2})^2$ state variables? The algorithm $\mathtt{verify}_A$ performs at most $c_2 n^{d_2}$ primitive operations and, assuming a computational model in which one bit of memory can be accessed per operation, references at most $c_2 n^{d_2}$ bits of memory. Its entire computation can therefore be summarized (more or less) using a $c_2 n^{d_2} \times c_2 n^{d_2}$ table, with rows corresponding to steps i and columns to bits j of memory. Each state variable then encodes the content of one bit of memory at one point in the computation.[11]

The consistency constraints sound complicated. But because one step of an algorithm (such as a Turing machine) is so simple, each of these logical constraints can be implemented with a small number of three-literal disjunctions (with the details depending on the precise computational model). The end result is a 3-SAT instance I_{3SAT} with a polynomial (in n) number of variables and constraints.

Postprocessor.   If the assumed subroutine returns a satisfying truth assignment for the 3-SAT instance I_{3SAT} constructed by the preprocessor, the postprocessor returns the candidate solution for I_A encoded

---

[11]There are additional details here that depend on the exact computational model used and the definition of a "primitive operation." The simplest approach is to use a Turing machine (see footnote 3), in which case another batch of Boolean variables is needed at each step to encode the machine's current internal state. The proof can be made to work for any reasonable model of computation.

by the assignments to the solution variables. If the assumed subroutine responds that I_{3SAT} is unsatisfiable, the postprocessor reports that I_A has no feasible solutions.

Outline of correctness. The constraints of the fabricated 3-SAT instance I_{3SAT} are defined so that the satisfying truth assignments correspond to the feasible solutions to the given instance I_A (encoded by the solution variables) along with the supporting verification performed by the algorithm $\texttt{verify}_A$ (encoded by the state variables). Thus, if I_A has no feasible solutions, the instance I_{3SAT} must be unsatisfiable. Conversely, if I_A does have a feasible solution, I_{3SAT} must have a satisfying truth assignment. The assumed 3-SAT subroutine must then compute such an assignment, which will be converted by the postprocessor into a feasible solution for I_A.

23.3.6   Solution to Quiz 23.1

Correct answer: (c). The exception? The influence maximization problem. While computing the total cost of a given tour or the makespan of a given schedule is straightforward, the influence of a given subset of k vertices is defined as an expectation with exponentially many terms (see (20.10) and the solution to Quiz 20.6). Because it's unclear how to evaluate the objective function in the influence maximization problem in polynomial time, the search version of the problem does not obviously belong to $\mathcal{NP}$.

*23.4   The P $\neq$ NP Conjecture

Way back in Section 19.3.5, we informally defined the P $\neq$ NP conjecture as: Checking an alleged solution to a problem can be fundamentally easier than coming up with your own solution from scratch. We are now, finally, in a position to state this conjecture formally.

23.4.1 $\mathcal{P}$: Polynomial-Time Solvable $\mathcal{NP}$ Problems

At least some of the problems in $\mathcal{NP}$ can be solved in polynomial time, such as the 2-SAT problem (Problem 21.12) and the search version of the minimum spanning tree problem (Section 19.1.1). The complexity class $\mathcal{P}$ is defined as the set of all such problems.

The Complexity Class $\mathcal{P}$

A search problem belongs to the complexity class $\mathcal{P}$ if and only if it belongs to $\mathcal{NP}$ and can be solved with a polynomial-time algorithm.

By definition, every problem in $\mathcal{P}$ also belongs to $\mathcal{NP}$:

$$\mathcal{P} \subseteq \mathcal{NP}.$$

23.4.2   Formal Definition of the Conjecture

No prizes for now guessing the formal statement of the P ≠ NP conjecture, that $\mathcal{P}$ is a strict subset of $\mathcal{NP}$:

The P ≠ NP Conjecture (Formal Version)

$$\mathcal{P} \subsetneq \mathcal{NP}.$$

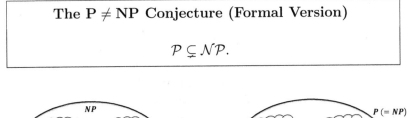

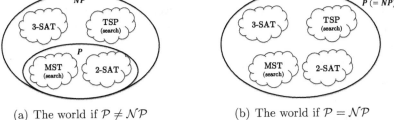

(a) The world if $\mathcal{P} \neq \mathcal{NP}$ (b) The world if $\mathcal{P} = \mathcal{NP}$

The P ≠ NP conjecture asserts the existence of a search problem with efficiently recognizable solutions (an $\mathcal{NP}$ problem) that cannot be solved by any polynomial-time algorithm—a problem for which checking an alleged feasible solution is easy, but coming up with your own from scratch is hard. If the conjecture is false, $\mathcal{P} = \mathcal{NP}$ and the efficient recognition of feasible solutions leads automatically to the efficient computation of feasible solutions (whenever they exist).

A polynomial-time algorithm for an NP-hard problem A would directly lead to one for every $\mathcal{NP}$ problem—with every problem in $\mathcal{NP}$ reducing to A, tractability would spread from A to all of $\mathcal{NP}$—proving that $\mathcal{P} = \mathcal{NP}$ and refuting the P ≠ NP conjecture. Thus, the

provisional definition of an NP-hard problem in Section 19.3.7 is a
logical consequence of the formal definition on page 189:

Consequence of NP-Hardness

If the P $\neq$ NP conjecture is true, no NP-hard problem can
be solved by a polynomial-time algorithm.

23.4.3   Status of the P $\neq$ NP Conjecture

The P $\neq$ NP conjecture is arguably the most important open question
in all of computer science, and also one of the deepest unsolved
problems in mathematics. For example, resolving the conjecture is
one of the seven "Millennium Problems" proposed in the year 2000
by the Clay Mathematics Institute; solve one of these problems and
you'll earn a prize of one million US dollars.[12,13]

Most experts believe that the P $\neq$ NP conjecture is true. (Although the legendary logician Kurt Gödel conjectured, in a 1956
letter to the still more legendary John von Neumann, a statement
equivalent to $\mathcal{P} = \mathcal{NP}$.) Why? For starters, humans are crafty at
discovering fast algorithms. If it really were the case that every single
problem in $\mathcal{NP}$ can be solved by a fast algorithm, why hasn't some
super-smart engineer or scientist discovered one yet? Meanwhile,
proofs delineating the limitations of algorithms have been few and far
between; if $\mathcal{P} \subsetneq \mathcal{NP}$, it's no real surprise that we haven't yet figured
out how to prove it.

Second, how could we reconcile $\mathcal{P} = \mathcal{NP}$ with the way the world
seems to work? We all "know," from direct experience, tasks for which
checking someone else's work (like a mathematical proof) takes far
less time and creativity than searching a large space of candidates for
your own solution. Yet $\mathcal{P} = \mathcal{NP}$ would imply that such creativity can
be efficiently automated; for example, at least in principle, a proof of

---

[12]The other six: the Riemann Hypothesis, the Navier-Stokes Equation, the
Poincaré Conjecture, the Hodge Conjecture, the Birch and Swinnerton-Dyer
Conjecture, and the Yang-Mills Existence and Mass Gap Problem. As of this
writing (in 2020), only the Poincaré Conjecture has been resolved (by Grigori
Perelman, in 2006, who famously refused the prize money).

[13]While one million dollars is nothing to sneeze at, it undersells the importance
and value of the advancement in human knowledge that appears necessary to
resolve a problem like the P $\neq$ NP conjecture.

Fermat's Last Theorem could then be generated by an algorithm in time polynomial in the length of the proof![14]

As a mathematical statement, how about some mathematical evidence for or against the P $\neq$ NP conjecture? Here, we know shockingly little. It may seem bizarre that no one has been able to prove such a seemingly obvious statement. The intimidating barrier is the dizzyingly rich fauna of the land of polynomial-time algorithms. If the "obvious" cubic running time lower bound for matrix multiplication is false (as shown by Strassen's algorithm in *Part 1*), who's to say that some other exotic species can't break through other "obvious" lower bounds, including the presumed ones for NP-hard problems?

Who's right: Gödel or Edmonds? You'd hope that, as the years go by, we'd be getting closer to a resolution of the P $\neq$ NP conjecture, one way or the other. Instead, as more and more mathematical approaches to the problem have proved inadequate, the solution has been receding further into the distance. We have to face the reality that we may not know the answer for a long time—certainly years, probably decades, and maybe even centuries.[15]

*23.5   The Exponential Time Hypothesis

23.5.1   Do NP-Hard Problems Require Exponential Time?

NP-hard problems are commonly conflated with problems that require exponential time to solve in the worst case (the third acceptable inaccuracy in Section 19.6). The P $\neq$ NP conjecture does not assert

---

[14]The ramifications of $\mathcal{P} = \mathcal{NP}$ would depend on whether all $\mathcal{NP}$ problems can be solved by algorithms that are fast in practice, or merely by algorithms that technically run in polynomial time but are too slow or complicated to be implemented and used. The first and more implausible scenario would have tremendous consequences for society, including the end of cryptography and modern ecommerce as we know it (see footnote 32 in Chapter 19); for a general-audience account of this possibility, see Lance Fortnow's book *A Golden Ticket* (Princeton University Press, 2013). The second scenario, which Donald E. Knuth himself has speculated about, would not necessarily have any practical implications; instead, it would signal that the mathematical definition of polynomial-time solvability is too liberal to accurately capture what we mean by "solvable by a fast algorithm in the physical world."

[15]For much more on the conjecture's broader context and current status, check out Scott Aaronson's book chapter "P $\overset{?}{=}$ NP" in *Open Problems in Mathematics*, edited by John F. Nash, Jr. and Michael Th. Rassias (Springer, 2016).

this, however, and even if true, leaves open the possibility that an NP-hard problem like the TSP could be solved in $n^{O(\log n)}$ or $2^{O(\sqrt{n})}$ time on instances with n vertices. The widely held belief that typical NP-hard problems require exponential time is codified by the *Exponential Time Hypothesis (ETH)*.[16]

The Exponential Time Hypothesis (ETH)

There is a constant $c > 1$ such that: Every algorithm that solves the 3-SAT problem requires time at least c^n in the worst case, where n denotes the number of variables.

The ETH does not preclude algorithms for the 3-SAT problem that improve over exhaustive search (which runs in time scaling with 2^n), and this is no accident: Problem 21.13 shows that there *are* much faster (if still exponential-time) algorithms for the problem. However, all the known 3-SAT algorithms require time c^n for some $c > 1$ (with the current record being $c \approx 1.308$); the ETH conjectures that this is unavoidable.

Reductions can be used to show that, if the ETH is true, many other natural NP-hard problems also require exponential time. For example, the ETH would imply that there is a constant $a > 1$ such that every algorithm for one of the NP-hard graph problems in Chapter 22 requires time at least a^n in the worst case, where n denotes the number of vertices.

23.5.2   The Strong ETH (SETH)

The Exponential Time Hypothesis is a stronger assumption than the $P \neq NP$ conjecture: If the former is true, so is the latter. Stronger assumptions lead to stronger conclusions; unfortunately, they are also more likely to be false (Figure 23.1)! Still, most experts believe the ETH is true.

Next is an even stronger assumption that is more controversial but has remarkable algorithmic implications. What could be stronger than assuming that an NP-hard problem requires exponential time to

---

[16]The stronger statement that *every* NP-hard problem requires exponential time is false; see Problem 23.5 for a contrived NP-hard problem that can be solved in subexponential time.

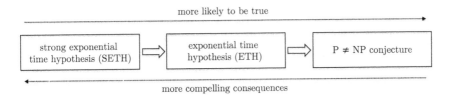

more likely to be true

more compelling consequences

Figure 23.1: Three unproven conjectures about the computational intractability of NP-hard problems, ordered from strongest (and least plausible) to weakest (and most plausible).

solve? *Assuming that there is no algorithm for the problem significantly faster than exhaustive search.* This can't possibly be true for the 3-SAT problem (on account of Problem 21.13); perhaps for a different problem? We don't need to look far—the general SAT problem, with no restriction on the number of disjunctions per constraint (page 136), is a plausible candidate.

The generality (and hence difficulty) of the k-SAT problem is nondecreasing in k, the maximum number of literals per constraint. Does the problem's difficulty strictly increase with k? For example, the randomized 3-SAT algorithm in Problem 21.13 can be extended to the k-SAT problem for every positive integer k, but its running time degrades with k as roughly $(2 - \frac{2}{k})^n$, where n is the number of variables (see footnote 44 on page 146). This same degradation in running time to 2^n as k increases shows up in all known k-SAT algorithms. Could it be necessary?

The Strong Exponential Time Hypothesis (SETH)

For every constant $c < 2$, there exists a positive integer k such that: Every algorithm that solves the k-SAT problem requires time at least c^n in the worst case, where n denotes the number of variables.[17]

Refuting the SETH would entail a major theoretical advance in satisfiability algorithms—a family of k-SAT algorithms (one per positive integer k), all of which run in $O((2 - \epsilon)^n)$ time, where n denotes the number of variables and $\epsilon > 0$ is a constant (independent of k and n,

---

[17]While not obvious, the SETH does imply the ETH.

like .01 or .001). Such an advance might or might not be in our near future, and expert opinion on the SETH is split; in any case, everyone is prepared for it to be refuted at any time.[18]

23.5.3 Running Time Lower Bounds for Easy Problems

Why tell you about a conjecture so strong that it might well be false? Because the SETH, a conjecture about the intractability of NP-hard problems, also has striking algorithmic implications for polynomial-time solvable problems.[19]

From the SETH to Sequence Alignment

In *Parts 1–3* of this book series we aspired to ever-faster algorithms, with the holy grail a blazingly fast linear- or near-linear-time algorithm. We achieved this goal for a number of problems (especially in *Parts 1 and 2*); for others (especially in *Part 3*), we stopped short. For example, for the sequence alignment problem (see footnote 24 on page 25 and Chapter 17 of *Part 3*), we declared victory with the NW (Needleman-Wunsch) dynamic programming algorithm that runs in $O(n^2)$ time, where n denotes the length of the longer of the two input strings.

Can we do better than this quadratic-time sequence alignment algorithm? Or, how would we amass evidence that we can't? The theory of NP-hardness, developed to reason about problems that seem unsolvable in polynomial time, appears irrelevant to this question. But a relatively new area of computational complexity theory, called *fine-grained complexity*, shows that hardness assumptions for NP-hard problems (like the SETH) translate meaningfully to polynomial-time solvable problems.[20] For example, a better-than-quadratic-time algorithm for the sequence alignment problem would automatically lead

---

[18]The ETH and SETH were formulated by Russell Impagliazzo and Ramamohan Paturi in their paper "On the Complexity of k-SAT" (*Journal of Computer and System Sciences*, 2001).

[19]The ETH also has some interesting algorithmic consequences that are not known to follow from the P $\neq$ NP conjecture. For example, if the ETH is true, many NP-hard problems and parameter choices do not allow for fixed-parameter algorithms (see footnote 22 on page 125).

[20]For a deep dive, check out the survey "On Some Fine-Grained Questions in Algorithms and Complexity," by Virginia Vassilevska Williams (*Proceedings of the International Congress of Mathematicians*, 2018).

to a better-than-exhaustive-search algorithm for the k-SAT problem for all k![21]

Fact 23.1 (SETH Implies NW Is Essentially Optimal) *For every constant $\epsilon > 0$, an $O(n^{2-\epsilon})$-time algorithm for the sequence alignment problem, where n is the length of the longer input string, would refute the SETH.*

In other words, the *only avenue* for improving on the running time of the NW algorithm is to make major progress on the SAT problem! This is a stunning connection between two problems that seem wildly different.

Reductions with Exponential Blow-Up

Fact 23.1, like all our NP-hardness proofs in Chapter 22, boils down to a reduction—actually, one reduction for each positive integer k—with k-SAT playing the role of the known hard problem and sequence alignment the role of the target problem:

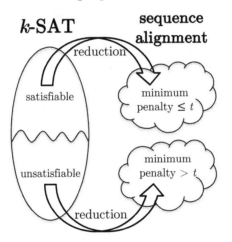

But how can we reduce an NP-hard problem to a polynomial-time solvable one without refuting the P $\neq$ NP conjecture? Each reduction behind Fact 23.1 employs a preprocessor that translates a k-SAT instance with n variables into a sequence alignment instance that is

---

[21]This result appears in the paper "Edit Distance Cannot Be Computed in Strongly Subquadratic Time (Unless SETH Is False)," by Arturs Backurs and Piotr Indyk (*SIAM Journal on Computing*, 2018).

exponentially bigger, with input strings that have length N in the ballpark of $2^{n/2}$.[22] The reduction also ensures that the fabricated instance has an alignment with total penalty at most a fabricated target t if and only if the given k-SAT instance is satisfiable, and that a postprocessor can easily extract a satisfiable assignment from an alignment with total penalty at most t.

Why a blow-up of $N \approx 2^{n/2}$? Because that number matches up the running times of the state-of-the-art algorithms for sequence alignment (dynamic programming) and k-SAT (exhaustive search). Composing such a reduction with an $O(N^2)$-time sequence alignment algorithm leads only to a k-SAT algorithm with running time $\approx 2^n$, the same as exhaustive search. By the same reasoning, a hypothetical $O(N^{1.99})$-time (say) sequence alignment subroutine would lead automatically (for every k) to an algorithm that solves the k-SAT problem in time roughly $O((2^{n/2})^{1.99}) = O((1.9931)^n)$. Because the base of this exponent is less than 2 (for all k), such an algorithm would refute the SETH.[23]

*23.6 NP-Completeness

A polynomial-time subroutine for an NP-hard problem like 3-SAT is all you need to solve every problem in $\mathcal{NP}$—every problem with efficiently recognizable solutions—in polynomial time. But something even stronger is true: Every problem in $\mathcal{NP}$ is *literally just a thinly disguised special case of 3-SAT*. In other words, the 3-SAT problem is *universal* among $\mathcal{NP}$ problems, in that it simultaneously encodes every single problem of $\mathcal{NP}$! This is the meaning of "NP-completeness." The search versions of almost all the problems studied in Chapter 22 are also NP-complete in this sense.

23.6.1 Levin Reductions

The idea that one search problem A is a "thinly disguised special case" of another search problem B is expressed through a highly

---

[22]This exponential blow-up evokes the exponentially large numbers essential to our reduction from the independent set problem to the subset sum problem (Theorem 22.9); see footnote 14 on page 173.

[23]Problem 23.7 outlines a simpler reduction of this type for the problem of computing the diameter of a graph.

restricted type of reduction known as a *Levin reduction*. Like the "simplest-imaginable reductions" introduced in Section 22.4, a Levin reduction carries out only the three unavoidable steps: transform in a preprocessing step a given instance of A to one of B; invoke the assumed subroutine for B; and transform in a postprocessing step the feasible solution returned by the subroutine (if any) into one for the given instance of A:[24,25]

algorithm for problem A

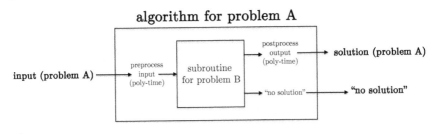

Levin Reduction from A to B

1. *Preprocessor:* Given an instance I of problem A, transform it in polynomial time to an instance I' of problem B.

2. *Subroutine:* Invoke the assumed subroutine for B with input I'.

3. *Postprocessor (feasible case):* If the subroutine returns a feasible solution to I', transform it in polynomial time to a feasible solution to I.

4. *Postprocessor (infeasible case):* If the subroutine returns "no solution," return "no solution."

Throughout this book, we've inadvertently used only Levin reductions and not the full power of general (Cook) reductions. Our

---

[24]Levin reductions conform to the template in Section 22.4 and, in addition: (i) both problems are required to be search problems; and (ii) the postprocessor is required to respond "no solution" if and only if the assumed subroutine does.

[25]If A and B are decision ("yes"/"no") problems rather than search problems, no postprocessing is necessary and the (binary) answer returned by the subroutine for B can be passed along unchanged as the final output. This analog of a Levin reduction for decision problems has a number of names: Karp reduction; polynomial-time many-to-one reduction; and polynomial-time mapping reduction.

reduction in Theorem 22.4 from the 3-SAT problem to the directed Hamiltonian path problem (page 167) is a canonical example: Given a 3-SAT instance, the preprocessor constructs a directed graph, which is then fed into a subroutine for computing an s-t Hamiltonian path, and if the subroutine returns such a path, the postprocessor extracts from it a satisfying truth assignment.[26]

23.6.2   The Hardest Problems in $\mathcal{NP}$

A problem B is NP-hard if it is algorithmically sufficient to solve all $\mathcal{NP}$ problems in polynomial time, meaning that for every problem A in $\mathcal{NP}$ there is a (Cook) reduction from A to B (page 189). To qualify as "NP-complete," a problem B must also belong to the class $\mathcal{NP}$ and include all other $\mathcal{NP}$ problems as thinly disguised special cases.

NP-Complete Problem

A computational problem B is *NP-complete* if:

1. For every problem A in $\mathcal{NP}$, there is a Levin reduction from A to B;

2. B is a member of the class $\mathcal{NP}$.

Because a Levin reduction is a special case of a (Cook) reduction, every NP-complete problem is automatically NP-hard.[27]

Of all the problems in $\mathcal{NP}$, the NP-complete problems are the hardest ones. Each such problem simultaneously encodes all search problems with efficiently recognizable solutions.[28]

---

[26]The other three main reductions in Chapter 22 (Theorems 22.2, 22.7, and 22.9) turn into Levin reductions once the optimization problem in the original is replaced by its search version (as you should check). For example, the reduction from the undirected Hamiltonian path problem to the TSP (Theorem 22.7) requires only a subroutine for the search version of the TSP (to check whether there is a zero-cost tour).

[27]Because of the second condition, only search problems can qualify for NP-completeness. For example, the TSP is NP-hard but not NP-complete, while its search version turns out to be both NP-hard *and* NP-complete.

[28]Most books define NP-completeness using decision (rather than search) problems and Karp (rather than Levin) reductions (footnote 25). The interpretation and algorithmic implications of NP-completeness are the same either way.

23.6.3   Existence of NP-Complete Problems

How cool is the definition of an NP-complete problem? A *single* search problem with efficiently recognizable solutions that simultaneously encodes *all* such search problems? It's amazing that such a problem could even exist!

But wait... we haven't actually seen any examples of NP-complete problems. Are there any? Could there really be such a "universal" search problem? Yes, and the Cook-Levin theorem already proves it! The reason is that its proof (Section 23.3.5) uses only a Levin reduction—a preprocessor that transforms instances of an arbitrary $\mathcal{NP}$ problem into 3-SAT instances and a postprocessor that extracts feasible solutions from satisfying truth assignments. Because the 3-SAT problem is also a member of $\mathcal{NP}$, it passes both tests for NP-completeness with flying colors.

Theorem 23.2 (Cook-Levin Theorem (Stronger Version))
The 3-SAT problem is NP-complete.

Proving from scratch that a problem is NP-complete is a tough task—Cook and Levin weren't awarded major prizes for nothing—but there's no need to do it more than once. Just as (Cook) reductions spread NP-hardness from one problem to another, Levin reductions spread NP-completeness (Problem 23.4):

Thus, to prove that a problem is NP-complete, just follow the three-step recipe (with the third step a check that the problem indeed belongs to $\mathcal{NP}$):

How to Prove a Problem Is NP-Complete

To prove that a problem B is NP-complete:

1. Prove that B is a member of the class $\mathcal{NP}$.

> 2. Choose an NP-complete problem A.
>
> 3. Prove that there is a Levin reduction from A to B.

This recipe has been applied many times over, and as a result we now know that thousands of natural problems are NP-complete, including problems from all across engineering, the life sciences, and the social sciences. For example, the search versions of almost all the problems studied in Chapter 22 are NP-complete (Problem 23.3).[29] The classic book by Garey and Johnson (see footnote 8 on page 156) lists hundreds more.[30]

The Upshot

☆ To amass evidence of a problem's intractability, prove that many other seemingly difficult problems reduce to it.

☆ In a search problem, the goal is to output a feasible solution or deduce that none exist.

☆ $\mathcal{NP}$ is the set of all search problems for which feasible solutions have polynomial length and can be verified in polynomial time.

☆ A problem is NP-hard if every problem in $\mathcal{NP}$ reduces to it.

---

[29]The exception? The influence maximization problem, the search version of which is not obviously in $\mathcal{NP}$ (see the solution to Quiz 23.1).

[30]The hopelessly inscrutable term "NP-complete" does a disservice to the fundamental concept that it defines, which deserves widespread appreciation and wonder. Lots of thought went into the name, however, as documented in Donald E. Knuth's article "A Terminological Proposal" (*SIGACT News*, 1974). Knuth's initial suggestions for what would become "NP-complete": "Herculean," "formidable," and "arduous." Write-in suggestions included "hard-boiled" (by Kenneth Steiglitz, as a hat tip to Cook) and "hard-ass" (by Albert R. Meyer, allegedly abbreviating "hard as satisfiability"). Meanwhile, Shen Lin suggested "PET" as a pleasingly flexible acronym, alternatively standing for: "probably exponential time," as long as the $P \neq NP$ conjecture is unresolved; "provably exponential time," if the conjecture is proved; and "previously exponential time," if the conjecture is refuted. (Now is not the time to nitpick and bring up Problem 23.5...)

☆ $\mathcal{P}$ is the set of all $\mathcal{NP}$ problems that can be solved with a polynomial-time algorithm.

☆ The P $\neq$ NP conjecture asserts that $\mathcal{P} \subsetneq \mathcal{NP}$.

☆ The exponential time hypothesis (ETH) asserts that natural NP-hard problems like the 3-SAT problem require exponential time.

☆ The strong exponential time hypothesis (SETH) asserts that, as k grows large, no algorithm for the k-SAT problem improves significantly over exhaustive search.

☆ If the SETH is true, no algorithm for the sequence alignment problem improves significantly over the Needleman-Wunsch algorithm.

☆ A Levin reduction carries out the minimum-imaginable work: preprocess the input; invoke the assumed subroutine; postprocess the output.

☆ A problem B is NP-complete if it belongs to the class $\mathcal{NP}$ and, for every problem $A \in \mathcal{NP}$, there is a Levin reduction from A to B.

☆ To prove that a problem B is NP-complete, follow the three-step recipe: (i) prove that $B \in \mathcal{NP}$; (ii) choose an NP-complete problem A; and (iii) design a Levin reduction from A to B.

☆ The Cook-Levin theorem proves that the 3-SAT problem is NP-complete.

Test Your Understanding

Problem 23.1 *(S)* Which of the following statements could be true, given the current state of knowledge? (Choose all that apply.)

a) There is an NP-hard problem that is polynomial-time solvable.

b) The $P \neq NP$ conjecture is true and also the 3-SAT problem can be solved in $2^{O(\sqrt{n})}$ time, where n is the number of variables.

c) There is no NP-hard problem that can be solved in $2^{O(\sqrt{n})}$ time, where n is the size of the input.

d) Some NP-complete problems are polynomial-time solvable, and some are not polynomial-time solvable.

Problem 23.2 *(S)* Prove that Edmonds's 1967 conjecture that (the optimization version of) the TSP cannot be solved by any polynomial-time algorithm is equivalent to the $P \neq NP$ conjecture.

Problem 23.3 *(S)* Which of the eighteen reductions listed in Section 22.3.2 can be easily turned into Levin reductions between the search versions of the corresponding problems?

Challenge Problems

Problem 23.4 *(H)* This problem formally justifies the recipes on pages 150 and 203 for proving that a problem is NP-hard and NP-complete, respectively.

(a) Prove that if a problem A reduces to a problem B and B reduces to a problem C, then A reduces to C.

(b) Conclude that if an NP-hard problem reduces to a problem B, then B is also NP-hard. (Use the formal definition of NP-hardness on page 189.)

(c) Prove that if there are Levin reductions from a problem A to a problem B and from B to a problem C, then there is a Levin reduction from A to C.

(d) Conclude that if a problem B belongs to $\mathcal{NP}$ and there is a Levin reduction from an NP-complete problem to B, then B is also NP-complete.

Problem 23.5 *(S)* Call an instance of the 3-SAT problem *padded* if its list of constraints concludes with n^2 redundant copies of the single-literal constraint "x_1," where n denotes the number of Boolean variables and x_1 is the first of those variables.

In the *PADDED 3-SAT* problem, the input is the same as in the 3-SAT problem. If the given 3-SAT instance is not padded or is unsatisfiable, the goal is to return "no solution." Otherwise, the goal is to return a satisfying truth assignment for the (padded) instance.

(a) Prove that the PADDED 3-SAT problem is NP-hard (or even NP-complete).

(b) Prove that the PADDED 3-SAT problem can be solved in subexponential time, namely $2^{O(\sqrt{N})}$ time for size-N inputs.

Problem 23.6 *(H)* Assume that the Exponential Time Hypothesis (page 196) is true. Prove that there exists a problem in $\mathcal{NP}$ that is neither polynomial-time solvable nor NP-hard.[31]

Problem 23.7 *(H)* The *diameter* of an undirected graph $G = (V, E)$ is the maximum shortest-path distance between any two vertices: $\max_{v,w \in V} dist(v, w)$, where $dist(v, w)$ denotes the minimum number of edges in a v-w path of G (or $+\infty$, if no such path exists).

(a) Explain how to compute the diameter of a graph in $O(mn)$ time, where n and m denote the number of vertices and edges of G, respectively. (You can assume that n and m are at least 1.)

(b) Assume that the Strong Exponential Time Hypothesis (page 197) is true. Prove that, for every constant $\epsilon > 0$, there is no $O((mn)^{1-\epsilon})$-time algorithm for computing the diameter of a graph.

---

[31] A famous and harder-to-prove result known as Ladner's Theorem shows that the conclusion remains true assuming only the (weaker) P $\neq$ NP conjecture.

Chapter 24

Case Study: The FCC Incentive Auction

NP-hardness is not some purely academic concept—it really does govern the range of computationally feasible options when solving a real-world problem. This chapter details a recent illustration of the importance of NP-hardness, in the context of a high-stakes economic problem: the efficient reallocation of a scarce resource (wireless spectrum). The solution deployed by the U.S. government, known as the FCC Incentive Auction, drew on an amazingly wide swath of the algorithmic toolbox that you've learned in this book. As you read through its details, take the time to appreciate the mastery of algorithms you've acquired since we first struggled through Karatsuba multiplication and the MergeSort algorithm in Chapter 1 of *Part 1*—how what started as a cacophony of mysterious and unconnected tricks has resolved into a symphony of interlocking algorithm design techniques.[1]

24.1 Repurposing Wireless Spectrum

24.1.1 From Television to Mobile Phones

Television spread like wildfire over the United States in the 1950s. In those days, television programming was transmitted solely over the air by radio waves, sent from a station's transmitter and received by a television's antenna. To coordinate stations' transmissions and prevent interference between them, the Federal Communications Commission (FCC) divvied up the usable frequencies—the *spectrum*—

---

[1]To learn more about the FCC Incentive Auction from its lead designers—Kevin Leyton-Brown, Paul Milgrom, and Ilya Segal—dig into their paper "Economics and Computer Science of a Radio Spectrum Reallocation" (*Proceedings of the National Academy of Sciences*, 2017). For a deep dive into the connections between auctions and algorithms, check out my book *Twenty Lectures on Algorithmic Game Theory* (Cambridge University Press, 2016).

into 6-megahertz (MHz) blocks called *channels*. Different stations in the same city would then broadcast on different channels. For example, "channel 14" refers to the frequencies between 470 MHz and 476 MHz; "channel 15" the frequencies between 476 MHz to 482 MHz; and so on.[2]

You know what else travels by radio waves over the air? All the data exchanged by your mobile phone and the nearest base station. For example, if it's the year 2020 and Verizon Wireless is your carrier, chances are you've been downloading and uploading data using the frequencies 746–756 MHz and 777–787 MHz, respectively. To avoid interference, the part of the spectrum reserved for cellular data does not overlap with that reserved for terrestrial (that is, over-the-air) television.

Mobile and wireless data usage has been exploding throughout the 21st century, increasing by roughly an order of magnitude over the past five years alone. Transmitting more data requires more dedicated frequencies, and not all frequencies are useful for wireless communication. (For example, with limited power, very high frequencies can carry signals only short distances.) Spectrum is a scarce resource, and modern technology is hungry for as much as it can get.

Television may still be big, but terrestrial television is not. Roughly 85-90% of U.S. households rely exclusively on cable television (which requires no over-the-air spectrum at all) or satellite television (which uses much higher frequencies than typical wireless applications). Reserving the most valuable spectrum real estate for over-the-air television made sense in the mid-20th century; no longer in the early 21st.

24.1.2   A Recent Reallocation of Spectrum

At the time of this writing, a major reallocation of spectrum is almost complete. After July 13, 2020, there will no longer be any television stations anywhere in the U.S. broadcasting over the air on what had been the highest channels, the fourteen channels between 38 and 51 (614–698 MHz). Every station that had been broadcasting on one of

---

[2]The ultra high frequency (UHF) channels start at 470 MHz and go up from there in 6 MHz blocks. The very high frequency (VHF) channels use lower frequencies, 174–216 MHz (for channels 7–13) and 54–88 MHz (for channels 2–6, along with 4 MHz for miscellaneous uses like garage door openers).

these channels is either switching to a lower channel or ceasing all terrestrial transmissions (while possibly still broadcasting via cable and satellite television). Even some of the stations that were already broadcasting on channels below 38 are going off the air or migrating to different channels, to make room for their comrades dropping down from higher channels. All told, 175 stations are relinquishing their broadcasting licenses and roughly 1000 are switching channels.[3]

The liberated 84 MHz of spectrum has been reorganized and awarded to telecommunication companies like T-Mobile, Dish, and Comcast, which are expected to use it to build out a new generation of wireless networks in the coming years. (T-Mobile, for example, has already flipped the switch on its new nationwide 5G network.) Where there had been channels 38–51, there are now seven independent pairs of 5 MHz blocks. For example, the first pair comprises the frequencies 617–622 MHz (meant for downloading to a device) and 663–668 MHz (meant for uploading); the second 622–627 MHz and 668–673 MHz; and so on.[4]

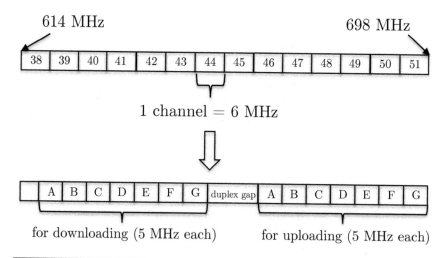

---

[3]Since the 2009 switchover from analog to exclusively digital broadcasting of terrestrial television, a logical channel (as displayed on a set-top box) can be remapped to a physical channel different from the one historically associated with that channel number. A station can therefore retain its logical channel even as its physical channel is reassigned.

[4]There's also an 11 MHz duplex gap (652–663 MHz) separating the two types of blocks, and a 3 MHz guard band (614–617 MHz) to avoid interference with channel 37 (608–614 MHz), which has long been reserved for radio astronomy and wireless medical telemetry.

This should all sound like a big, messy operation. Which stations should go off the air? Which should switch channels? What should their new channels be? How much should station owners be compensated for their losses? Which telecoms should be awarded the newly created paired blocks of spectrum? What should they pay for them? These are all questions answered by the *FCC Incentive Auction*—a complex algorithm that leaned heavily on the toolbox for tackling NP-hard problems described in this book.

24.2   Greedy Heuristics for Buying Back Licenses

The FCC Incentive Auction had two parts: a *reverse auction* for deciding which television stations would go off the air or switch channels, and the appropriate compensation for them; and a *forward auction* for choosing who receives blocks of the newly freed spectrum, and at what prices. The U.S. government (along with many other countries) has been running forward auctions to sell spectrum licenses with great success for twenty-five years, making small tweaks to them along the way. This case study focuses on the FCC Incentive Auction's unprecedented reverse auction, wherein lay most of its innovation.

24.2.1   Four Temporary Simplifying Assumptions

The FCC confers property rights to a television broadcaster through a *broadcasting license*, which authorizes broadcasting over a channel in a specified geographic region. The FCC assumes responsibility for ensuring that each station suffers little to no interference across its broadcast area.[5]

The goal of the reverse auction in the FCC Incentive Auction was to reclaim enough licenses from television stations to free up a target amount of spectrum (like channels 38–51). To get an initial feel for this problem, let's make some simplifying assumptions, to be removed as we go along:

---

[5]For the purposes of the FCC Incentive Auction, the specific channel assignment of a station was not considered part of the license owner's property rights. An act of Congress was required to authorize this interpretation and allow the auction to reassign stations' channels as needed. (One of only eight bills passed by Congress in 2012, perhaps because of its veto-proof title: the "Middle Class Tax Relief and Job Creation Act.")

> ## Temporary Simplifying Assumptions
>
> 1. All stations that remain on the air will broadcast on a single channel (channel 14, say).
>
> 2. Two stations can broadcast on the same channel simultaneously if and only if their broadcasting areas do not overlap.
>
> 3. There is a known value for each station.
>
> 4. The government can unilaterally decide which stations remain on the air.

Ideally, the most valuable stations would be the ones to retain their licenses. The objective, then, would be to identify a set of non-interfering stations with the maximum-possible sum of station values. Do you recognize this optimization problem?

24.2.2 Ambushed by Weighted Independent Set

It's exactly the weighted independent set problem (page 19)! Vertices correspond to stations, edges to pairs of interfering stations, and station values to vertex weights:

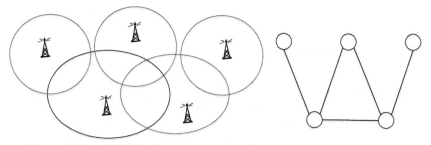

(five stations & their broadcast areas)   (corresponding graph)

We know from Corollary 22.3 that this problem is NP-hard, even when every vertex has weight 1. The problem can be solved in linear time using dynamic programming when the input graph is a tree (see Chapter 16 of *Part 3*), but the interference patterns of television stations are not at all tree-like. For example, all stations in the same city interfere with each other, leading to a clique in the corresponding graph.

Searching Through the Algorithmic Toolbox

Now that we have diagnosed the problem as NP-hard, it's time to search for a cure in the appropriate compartment of our algorithmic toolbox. (NP-hardness is not a death sentence!) Most ambitiously, could the problem be solved exactly in a tolerable amount of time—say, under a week?

The answer depends on the size of the problem. If only thirty stations were involved, exhaustive search would work just fine. But the real problem had thousands of participating stations and tens of thousands of interference constraints—well above the pay grade of exhaustive search and the dynamic programming techniques in Sections 21.1–21.2.

The last hope for an exact algorithm would be a semi-reliable magic box for optimization problems like a MIP solver (Section 21.4). The weighted independent set problem is easily encoded as a MIP problem (Problem 21.9), and this was exactly what the FCC tried first. Unfortunately, the problem proved too big, and even the latest and greatest MIP solvers choked on it. (Or at least, they choked on the more realistic multi-channel version of the problem described in Section 24.2.4.) With all options exhausted for a 100% correct algorithm, the FCC had no choice but to compromise on correctness and turn to fast heuristic algorithms.

24.2.3   Greedy Heuristic Algorithms

For the weighted independent set problem, as with so many others, greedy algorithms are the perfect place to start brainstorming about fast heuristic algorithms.

The Basic Greedy Algorithm

Perhaps the simplest greedy approach to the weighted independent set problem is to mimic Kruskal's minimum spanning tree algorithm and perform a single pass over the vertices (in decreasing order of weight), always adding a vertex to the output unless it destroys feasibility:

WISBasicGreedy

Input: undirected graph $G = (V, E)$ and a nonnegative weight w_v for each vertex $v \in V$.
Output: an independent set of G.

$S := \emptyset$
sort vertices of V from highest to lowest weight
`// Main loop`
for each $v \in V$, in nonincreasing order of weight **do**
    if $S \cup \{v\}$ is feasible **then**   `// all non-adjacent`
        $S := S \cup \{v\}$
return S

For example, in the graph

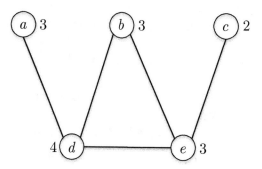

(vertices labeled with their weights)

the `WISBasicGreedy` algorithm selects the vertex d with the largest weight in its first iteration, skips the weight-3 vertices in its second, third, and fourth iterations (because each is adjacent to d), and concludes by selecting vertex c. The resulting independent set has total weight 6 and is not optimal (as the independent set $\{a, b, c\}$ has total weight 8).

Because the weighted independent set problem is NP-hard and the `WISBasicGreedy` algorithm runs in polynomial time, we were fully expecting examples of this type. But here's a more troubling case (with vertices labeled with their weights):

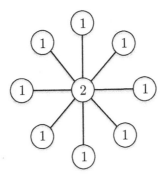

The `WISBasicGreedy` algorithm is tricked into committing to the center of the star, precluding it from taking any of the leaves. How can we discourage pitfalls of this type?

Vertex-Specific Multipliers

To avoid the miscues of the `WISBasicGreedy` algorithm, we can discriminate against vertices with many neighbors. For example, acknowledging that selecting a vertex v accrues benefit w_v while knocking out $1 + \deg(v)$ vertices from consideration (where $\deg(v)$ denotes v's degree), the algorithm's single pass could be in decreasing order of bang-per-buck $w_v/(1+\deg(v))$ rather than of weight w_v.[6] (This greedy algorithm returns the maximum-weight independent set in our two examples.) More generally, the algorithm can compute vertex-specific multipliers however it likes in a preprocessing step before proceeding to its single pass over vertices:

WISGeneralGreedy

compute β_v for each $v \in V$ // ex: $\beta_v = 1 + \deg(v)$
$S := \emptyset$
sort vertices of V from highest to lowest value of w_v/β_v
for each $v \in V$, in nonincreasing order of w_v/β_v **do**
    if $S \cup \{v\}$ is feasible **then**   // all non-adjacent
        $S := S \cup \{v\}$
return S

---

[6]You might recognize this idea from Problem 20.3 and the greedy heuristic algorithm for the knapsack problem that sorts items in decreasing order of value-size ratios.

What's the best choice for the vertex-specific multipliers? No matter how smart the formula for each parameter β_v, there will be examples in which the WISGeneralGreedy algorithm returns a suboptimal independent set (assuming that the β_v's can be computed in polynomial time and that the P $\neq$ NP conjecture is true). The best choice depends on the problem instances that tend to show up in the application of interest, and should therefore be determined empirically using representative instances.[7]

Station-Specific Parameters in the FCC Incentive Auction

Representative instances of the weighted independent set problem, and the more general multi-channel problem described in the next section, were easy to come by in the design phase of the reverse auction. The graph—derived from the participating stations and their broadcast areas—was fully known in advance. Educated guesses could be made about the range of likely vertex weights (station values) based on historical data. With a carefully tuned choice of vertex-specific multipliers, the WISGeneralGreedy algorithm (and the multi-channel generalization FCCGreedy described in the next section) routinely returned solutions to representative instances with total weight exceeding 90% of the maximum possible.[8,9]

24.2.4   The Multi-Channel Case

Time to discard the first simplifying assumption in Section 24.2.1 and allow still-on-air stations to be assigned any one of k channels.

---

[7]General advice for tackling NP-hard problems in a real application: Exploit as much domain-specific knowledge as you can!

[8]How were the parameters computed in the actual FCC Incentive Auction? Via the formula $\beta_v = \sqrt{\deg(v)} \cdot \sqrt{\mathrm{pop}(v)}$, where $\deg(v)$ and $\mathrm{pop}(v)$ denote the number of stations overlapping with and the population served by the station v, respectively. The $\sqrt{\deg(v)}$ term discriminated against stations that would block lots of other stations from remaining on the air. The point of the $\sqrt{\mathrm{pop}(v)}$ term was more subtle (and controversial); its effect was to decrease the compensation paid by the government to small television stations that were likely to go off the air anyway.

[9]The FCC was also able to obtain high-quality solutions in a reasonable amount of time by stopping a state-of-the-art MIP solver early, prior to finding an optimal solution (see page 133). The greedy approach ultimately won out on account of its easy translation to a transparent auction format (as detailed in Section 24.4).

The WISGeneralGreedy algorithm would seem to extend easily to the multi-channel version of the problem:[10]

```
                        FCCGreedy

compute βᵥ for each station v
S := ∅
sort stations from highest to lowest value of wᵥ/βᵥ
for each station v, in nonincreasing order of wᵥ/βᵥ do
    if S ∪ {v} is feasible then   // fit on k channels
        S := S ∪ {v}
return S
```

Looks like all our other (polynomial-time) greedy algorithms, right? But let's drill down on an iteration of the main loop, which is responsible for testing whether the current station v can be added to the solution-so-far S without destroying feasibility. What makes a subset of stations "feasible"? Feasibility means that the stations can all be on the air at the same time, without interference. That is, there should be an assignment of the stations in $S \cup \{v\}$ to the k available channels so that no two stations with overlapping broadcast areas are assigned the same channel. Do you recognize this computational problem?

24.2.5   Ambushed by Graph Coloring

It's exactly the graph coloring problem (page 135)! Vertices correspond to stations, edges to pairs of stations with overlapping broadcast areas, and the k colors to the k available channels.

As we know from Problem 22.11, the graph coloring problem is NP-hard even when $k = 3$.[11] Worse still, the FCCGreedy algorithm must solve *many* instances of the graph coloring problem, one in each iteration of its main loop. How are these instances related?

---

[10]Think of k as 23, corresponding to channels 14–36. The FCC Incentive Auction also allowed UHF stations to drop down to the VHF band (channels 2–13), but most of the action took place in the UHF band.

[11]Checking feasibility in the special case of a single channel (Section 24.2.2) corresponds to the trivial problem of checking 1-colorability or, equivalently, checking whether a set of vertices constitutes an independent set.

Quiz 24.1

Consider the sequence of feasibility-checking instances that arises in the FCCGreedy algorithm. Which of the following statements are true? (Choose all that apply.)

a) If the instance in one iteration is feasible, so is the instance in the next iteration.

b) If the instance in one iteration is infeasible, so is the instance in the next iteration.

c) The instance in a given iteration has one more station than the instance in the previous iteration.

d) The instance in a given iteration has one more station than the most recent feasible instance.

(See Section 24.2.6 for the solution and discussion.)

Now what? Does diagnosing our feasibility-checking problem as the NP-hard graph coloring problem rule out using a greedy heuristic algorithm to approximately maximize the total value of the stations that remain on the air?

24.2.6   Solution to Quiz 24.1

Correct answer: (d). Answer (a) is obviously incorrect: The first instance is always feasible, while some of the instances toward the end of the algorithm may not be. Answer (b) is also incorrect; for example, the solution-so-far might block all newcomers in the northeastern region of the U.S. while leaving its west coast wide open. Answer (c) is incorrect and (d) is correct, as the solution-so-far S changes only in an iteration in which the station set $S \cup \{v\}$ is feasible. For example, phrased in terms of graph coloring:

[with $k=2$]

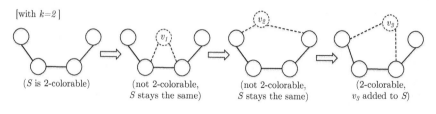

| (S is 2-colorable) | (not 2-colorable, S stays the same) | (not 2-colorable, S stays the same) | (2-colorable, v_3 added to S) |

24.3   Feasibility Checking

If only we had a magic box for checking feasibility, we could run the FCCGreedy algorithm and, hopefully, after carefully tuning the station-specific multipliers, reliably compute feasible solutions with total value close to the maximum possible. Our dreams of magic boxes have already been thwarted once, with the original value-maximization problem proving too tough for the latest and greatest MIP solvers (Section 24.2.2). Why should we expect any more success this time around?

24.3.1   Encoding as a Satisfiability Problem

The subroutine required by the FCCGreedy algorithm is responsible only for feasibility checking (corresponding to checking whether a given subgraph is k-colorable), not optimization (corresponding to finding the maximum-value k-colorable subgraph of a given graph). This raises the hope for a magic box that solves the easier (if still NP-hard) feasibility-checking problem, even if none exist for the optimization problem. The pivot from optimization to feasibility checking also suggests experimenting with a different language and technology—logic and SAT solvers, rather than arithmetic and MIP solvers.

The formulation of the graph coloring problem as a satisfiability problem in Section 21.5.3 is immediately relevant here. To review, for each vertex v in the input graph and allowable color $i \in \{1, 2, \ldots, k\}$, there is a Boolean (true/false) variable x_{vi}. For each edge (u, v) of the input graph and color i, there is a constraint

$$\neg x_{ui} \lor \neg x_{vi} \tag{24.1}$$

that rules out assigning the color i to both u and v. For each vertex v of the input graph, there is a constraint

$$x_{v1} \lor x_{v2} \lor \cdots \lor x_{vk} \tag{24.2}$$

that rules out leaving v colorless.

Optionally, for each vertex v and distinct colors $i, j \in \{1, 2, \ldots, k\}$, the constraint

$$\neg x_{vi} \lor \neg x_{vj} \tag{24.3}$$

can be used to rule out assigning both color i and color j to v.[12]

24.3.2   Incorporating Side Constraints

The actual FCC Incentive Auction used a formulation slightly more complicated than (24.1)–(24.3). Stations with overlapping broadcast areas interfere when assigned the same channel and, depending on several factors, may also interfere when assigned adjacent channels (like 14 and 15). A separate team at the FCC determined in advance, for each pair of stations, exactly which pairs of channel assignments would create interference. This list of forbidden pairwise channel assignments, while difficult to compile, was straightforward to incorporate into the satisfiability formulation, with one constraint of the form

$$\neg x_{uc} \vee \neg x_{vc'} \tag{24.4}$$

for each pair u, v of stations and forbidden channel assignments c, c' to them. For example, the constraint $\neg x_{u14} \vee \neg x_{v15}$ would prevent the stations u and v from being assigned channels 14 and 15, respectively. This list of interference constraints replaces the second of the simplifying assumptions in Section 24.2.1.

Another wrinkle was that not all stations were eligible for all channel assignments. For example, stations that bordered Mexico could not be assigned to a channel that would interfere with an existing station on the Mexican side of the border. To reflect these additional constraints, the decision variable x_{vi} was omitted whenever the station v was forbidden from channel i.

These tweaks to the original SAT formulation (24.1)–(24.3) illustrate a general strength of MIP and SAT solvers, relative to problem-specific algorithm design: They are often better at accommodating all kinds of idiosyncratic side constraints with minimal modifications to the basic formulation.

24.3.3   The Repacking Problem

The feasibility-checking problem in the reverse auction of the FCC Incentive Auction was almost but not quite a graph coloring problem

---

[12]Vertices can receive multiple colors if these constraints are omitted, but every way of choosing among the assigned colors results in a k-coloring.

(because of the side constraints in Section 24.3.2), so let's give it a new name: the *repacking problem*.

Problem: The Repacking Problem

Known in advance: A list V of television stations, the allowable channels C_v for each station $v \in V$, and the allowable channel pairs P_{uv} for each station pair $u, v \in V$.

Input: A subset $S \subseteq V$ of television stations.

Output: An assignment of each station $v \in S$ to a channel in C_v such that each station pair $u, v \in S$ is assigned a channel pair in P_{uv}. (Or, correctly declare that no such assignment exists.)

Call a subset of stations *packable* if the corresponding repacking instance has a feasible solution, and *unpackable* otherwise.

The FCC's algorithmic aspirations were ambitious: to solve the repacking problem reliably in a minute or less! (We'll see in Section 24.4 why the time budget was so small.) Repacking instances in the FCC Incentive Auction had thousands of stations, tens of thousands of pairs of overlapping stations, and dozens of available channels. After the translation to satisfiability (as in Sections 24.3.1–24.3.2), the resulting instances had tens of thousands of decision variables and more than one million constraints.

That's pretty big! Still, why not throw the latest and greatest SAT solvers at them and see how they do? Unfortunately, when applied off the shelf, these solvers frequently needed ten minutes or more to solve representative repacking instances. Doing better required throwing the kitchen sink at the problem.

24.3.4   Trick #1: Presolvers (Look for an Easy Way Out)

The FCC Incentive Auction used *presolvers* to quickly ferret out instances that were obviously packable or unpackable. These presolvers exploited the nested structure of the repacking instances in the FCCGreedy algorithm (see Quiz 24.1), with each instance taking the form $S \cup \{v\}$ for a packable set of stations S and a new station v.

For example, the auction administered two quick and dirty local tests that examined only the (relatively small) neighborhood of v. Formally, call two stations *neighbors* if they appear jointly in at least one interference constraint (24.4), and let $N \subseteq S$ denote the neighbors of v in S.

1. Check if $N \cup \{v\}$ is packable; if not, halt and report "unpackable." (Correctness: Supersets of unpackable sets of stations are themselves unpackable.)

The analog in a graph coloring instance would be to check if a given vertex v and its neighbors form a k-colorable subgraph. For example (with $k = 2$):

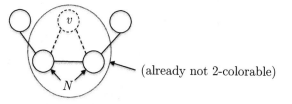

(already not 2-colorable)

2. Inherit the previously computed feasible channel assignments for the (packable) stations S. Hold the assignments of all stations in $S - N$ fixed. Check if there are channel assignments to the stations in $N \cup \{v\}$ so that the combined assignments are feasible. If so, report "packable" and return the combined channel assignments.

Whether this step succeeds generally depends on the inherited channel assignments for the stations in $S - N$. For example (with $k = 3$):

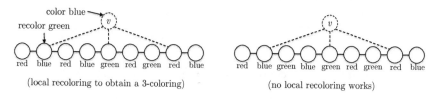

(local recoloring to obtain a 3-coloring)   (no local recoloring works)

The size of the neighborhood N was typically in the single or double digits, so each of these steps could be carried out quickly using a SAT solver. Ambiguity remained for the repacking instances that passed through both steps. Such an instance could have been packable, due to a feasible channel assignment deviating from the restricted form considered in step 2. Or it could have been unpackable, with no packing of $N \cup \{v\}$ in step 1 extendable to one of all of $S \cup \{v\}$.

24.3.5   Trick #2: Preprocess and Simplify

Every repacking instance that survived the presolvers was subjected to a preprocessing step designed to reduce its size.[13]

Removing Easy Stations

Call a station u of $S \cup \{v\}$ *easy* if, no matter what the other stations' channel assignments are, u can be assigned a channel of C_u that avoids interference with all of its neighbors. (The analog in a graph coloring instance would be a vertex whose degree is smaller than the number k of colors.)

3. Iteratively remove easy stations: (i) initialize $X := S \cup \{v\}$; (ii) while X contains an easy station u, $X := X - \{u\}$.

For example, in a graph coloring context (with $k = 3$):

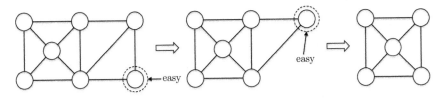

Quiz 24.2

Iteratively removing easy stations from a set $S \cup \{v\}$...

a) ... could change the set's status from packable to un-packable, or vice versa.

b) ... could change the set's status from packable to un-packable, but not from unpackable to packable.

---

[13]This idea is similar in spirit to the "for-free primitives" emphasized throughout this book series. If you have a blazingly fast primitive (like sorting, computing connected components, etc.) that might simplify your problem, why not use it?

c) ...could change the set's status from unpackable to packable, but not from packable to unpackable.

d) ...cannot change the set's packability status.

(See Section 24.3.8 for the solution and discussion.)

Decomposing the Problem

The next step sought to decompose the problem into smaller independent subproblems. (The analog in a graph coloring instance would be to compute a k-coloring separately for each connected component.)

4. Given a set of (non-easy) stations:

   a) Form a graph H with vertices corresponding to stations and edges to neighboring stations.

   b) Compute the connected components of H.

   c) For each connected component, solve the corresponding repacking problem.

   d) If at least one subproblem is unpackable, report "unpackable." Otherwise, report "packable" and return the union of the channel assignments computed in the subproblems.

Because stations interfere only with neighboring stations, the different subproblems do not interact in any way. The stations in X are therefore packable if and only if all the independent subproblems are packable.

Why did decomposing the problem help? The FCC Incentive Auction remained on the hook for solving all the subproblems, whose combined size was the same as that of the original problem. But whenever you have an algorithm that runs in super-linear time (as one would expect from a SAT solver), it's faster to solve an instance in pieces than all at once.[14]

---

[14]For example, consider a quadratic-time algorithm, running in time cn^2 on size-n instances for some constant $c > 0$. Solving two size-$(n/2)$ instances then takes $2 \cdot c(n/2)^2 = cn^2/2$ time, a factor-2 speedup over solving a single size-n instance.

24.3.6   Trick #3: A Portfolio of SAT Solvers

The toughest repacking instances survived the gauntlet of presolvers and preprocessing and awaited more sophisticated tools. While every state-of-the-art SAT solver had success on some representative instances, none met the FCC mandate of reliably solving instances in a minute or less. What next?

The designers of the reverse auction in the FCC Incentive Auction took advantage of two things: (i) the empirical observation that different SAT solvers struggle on different instances; and (ii) modern computer processors. Rather than putting all its eggs in one basket with a single SAT solver, the auction used a portfolio of *eight* carefully tuned solvers, running in parallel on an 8-core workstation.[15,16] This, finally, was sufficient algorithmic firepower to solve over 99% of the repacking instances faced by the auction within the target of one minute each. Pretty impressive for satisfiability instances with tens of thousands of variables and more than one million constraints!

24.3.7   Tolerating Failures

Over 99% sounds pretty good, but what happened the remaining 1% of the time? Did the FCC Incentive Auction spin its wheels helplessly while eight SAT solvers fumbled around, desperate for a satisfying assignment?

Another feature of the FCCGreedy algorithm is its tolerance of failures by its feasibility-checking subroutine. Suppose, when checking the feasibility of a set $S \cup \{v\}$, the subroutine times out and reports "I don't know." Without an assurance of feasibility (which is an ironclad constraint), the algorithm cannot risk adding v to its solution and must skip it, potentially foregoing some of the value it could have otherwise obtained. But the algorithm always finishes in a predictable

---

[15]And how were these eight solvers chosen? With a greedy heuristic algorithm analogous to those for the maximum coverage (Section 20.2) and influence maximization (Section 20.3) problems! The solvers were chosen sequentially, with each solver maximizing the marginal running time improvement on representative instances, relative to the solvers already in the portfolio.

[16]For fans of local search (Sections 20.4–20.5) distraught over its apparent absence from this case study: Several SAT solvers in this portfolio were local search algorithms—think greedier and highly parameterized versions of the randomized SAT algorithm described in Problem 21.13.

amount of time with a feasible solution, and the loss in value from timeouts should be modest provided they are infrequent, as they were in the FCC Incentive Auction.

24.3.8   Solution to Quiz 24.2

Correct answer: (d). If the final set X is unpackable, so is the superset $S \cup \{v\}$. If X is packable, every feasible channel assignment to the stations of X can be extended to all of $S \cup \{v\}$, one easy station at a time (in reverse order of removal):

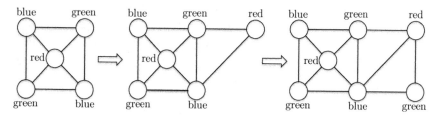

24.4 Implementation as a Descending Clock Auction

Where's the "auction" in the FCC Incentive Auction? Doesn't the FCCGreedy algorithm in Section 24.2, along with the repacking subroutine in Section 24.3, already solve the value-maximization problem to near-optimality? With at most a few thousand feasibility checks (one per participating station) and one minute spent per feasibility check, the algorithm would finish in a matter of days. Time to declare victory?

No. Time instead to revisit and remove the last two simplifying assumptions in Section 24.2.1. Stations were not forcibly removed from the air; they relinquished their licenses voluntarily (in exchange for compensation). So why not run the FCCGreedy algorithm to figure out which stations should stay on the air and buy out the other stations at whatever price they'd be willing to accept? Because the value of a station, defined here as the minimum compensation its owner would accept for going off the air, was not known in advance. (You could ask the owner, but they would probably overstate their value in the hopes of receiving extra compensation.) How could the FCCGreedy algorithm possibly be implemented without advance knowledge of stations' values?

24.4.1   Auctions and Algorithms

Think back to auctions that you've seen in the movies or in real life—perhaps at an estate sale, an auction house, or a school fundraiser. An auctioneer asks questions of the form "who's willing to buy this tennis ball signed by Roger Federer for one hundred bucks?" and the willing buyers raise their hands. In the reverse auction of the FCC Incentive Auction, the "auctioneer" (the government) was buying rather than selling, so the questions had the form "who's willing to sell their broadcasting license for one million dollars"? A station's response to this question with offered compensation p revealed whether its value—the minimum acceptable compensation—was above or below p.

The FCCGreedy algorithm begins by sorting stations in nonincreasing order of w_v/β_v, where w_v is the value of station v and β_v is a station-specific parameter—an apparent nonstarter when station values are unknown.[17] Can we reimplement the algorithm so that the stations effectively sort themselves, using only auction-friendly operations of the form "is $w_v \leq p$"?

24.4.2   Example

To see how this might work, assume for now that stations' values are positive integers between 1 and a known upper bound W. Assume also that there is only one free channel ($k = 1$) and that $\beta_v = 1$ for every station v. For example, suppose there are five stations and $W = 5$:

(stations labeled with their values)

---

[17]In the FCC Incentive Auction, the station-specific parameters β_v *were* known in advance, as they depended only on the population served by and interference constraints of a station (see footnote 8).

The idea is to start with the highest-imaginable compensation ($p = W$) and work downward. The set S of stations to remain on the air is initially empty. In the algorithm's first iteration, each station's value is compared to the initial value of p (that is, to 5); equivalently, each broadcaster is asked if they would accept a compensation of 5 in exchange for their license. All participants accept, and the algorithm decrements p and proceeds to the next iteration. All participants again accept the reduced compensation offer (with $p = 4$). The station with value 4 refuses the offer at $p = 3$ in the next iteration, and the algorithm responds by adding it to the on-air station set S:

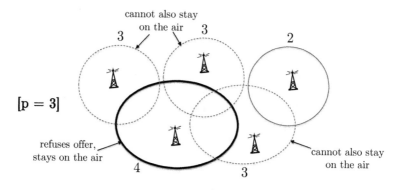

With the value-4 station now back on the air and only one channel available, the three overlapping stations are blocked and must stay off the air. In subsequent iterations, the algorithm makes decreasing offers of compensation to the only station whose fate remains unresolved, the value-2 station. That station refuses the offer at $p = 1$, at which point it is added to S and the algorithms halts:

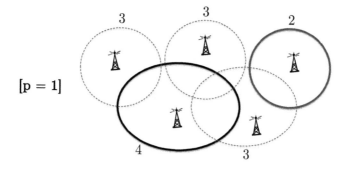

In this example and in general, this iterative process recreates the trajectory of the WISBasicGreedy algorithm on the corresponding weighted independent set instance (Section 24.2.3): Stations drop out (and go back on the air) in nonincreasing order of value, subject to feasibility. Can we extend this idea to capture the full-fledged FCCGreedy algorithm?

24.4.3   Reimplementing the FCCGreedy Algorithm

The station-specific parameters β_v in the FCCGreedy algorithm can be emulated using station-specific offers, with compensation $\beta_v \cdot p$ offered to station v in an iteration with "base price" p. A station v with value w_v will then drop out when the base price p drops below w_v/β_v. As p gradually decreases, the resulting process faithfully simulates the FCCGreedy algorithm: Stations drop out (and go back on the air) in nonincreasing order of w_v/β_v, subject to feasibility. The resulting algorithm is called a *descending clock auction,* and it is the exact one used in the FCC Incentive Auction's reverse auction (with $\epsilon = 0.05$ and the β_v's defined as in footnote 8):

FCCDescendingClock

Input: set V of stations, parameter $\beta_v > 0$ for each $v \in V$, parameter $\epsilon \in (0,1)$.
Output: a repackable subset $S \subseteq V$.

$p :=$ LARGE NUMBER `// maximize participation`
$S := \emptyset$ `// stations staying on the air`
$X := \emptyset$ `// stations going off the air`
while $S \cup X \neq V$ **do**   `// still stations in limbo`
   for each station $v \notin S \cup X$, in arbitrary order **do**
      `// invoke feasibility checker (§24.3)`
      if $S \cup \{v\}$ packable **then** `// still room for v`
         offer compensation $\beta_v \cdot p$ to v
         if offer refused **then** `// because` $p < w_v/\beta_v$
            $S := S \cup \{v\}$ `// v goes back on air`
      else   `// no room for v (or timeout)`
         $X := X \cup \{v\}$ `// v must stay off air`
   $p := (1-\epsilon) \cdot p$ `// lower offers in next round`
return S

The outer loop of the FCCDescendingClock algorithm controls the value p of the "clock." This base price decreases by a small amount in each iteration (called a *round*), until all stations' fates have been sealed. Within a round, the inner loop performs a single pass over the remaining stations in an arbitrary order, with the intention of making a new, lower offer of compensation to each. But before making a lower offer to a station v, the algorithm calls on the feasibility checker from Section 24.3 to ensure that v could be accommodated on the air should it decline the offer.[18] If the feasibility checker finds that $S \cup \{v\}$ is unpackable, or if it times out, the algorithm cannot risk a refusal by v and commits to keeping v off the air. If the feasibility checker comes back with a feasible packing of the stations in $S \cup \{v\}$, the algorithm can proceed safely with the lower offer to v. The algorithm returns the final set S of stations to remain on the air, along with the channel assignments computed for them by the feasibility checker.

24.4.4   Time to Get Paid

The FCCDescendingClock algorithm determines the stations remaining on the air and their new channel assignments. It has one more responsibility: to compute the prices paid to the departing broadcasters in exchange for their licenses. (Broadcasters that remain on the air receive no compensation.[19])

First, what's the initial base price p? In the FCC Incentive Auction, this value was chosen so that the opening offers would be absurdly lucrative and entice many stations to participate (participation was voluntary). For example, the opening offer to WCBS, the CBS affiliate in New York City, was 900 million dollars![20] Every broadcaster that entered the auction was contractually obligated to sell its license at

---

[18]The original FCCGreedy algorithm in Section 24.2.4 invokes the feasibility checker only once per participating station. The reimplemented version, the FCCDescendingClock algorithm, requires a new batch of feasibility checks *in every round*. The FCC Incentive Auction's reverse auction ran for dozens of rounds, requiring roughly one hundred thousand feasibility checks in all. This is why the FCC granted only one minute per feasibility check. (And even with one-minute timeouts, the auction took many months to complete.)

[19]Technically, stations forced to switch channels after the FCC Incentive Auction received a modest sum of money—far less than the typical selling price of a license—to cover the switching costs.

[20]And remember that selling a license only meant giving up on terrestrial broadcasting—small potatoes compared to cable and satellite television.

the opening offer should the government request it, and similarly for every subsequent (lower) offer accepted during the auction. The government, naturally, paid the lowest agreed-upon price:

> ### Compensation in the FCC Incentive Auction
>
> Each broadcaster going off the air was paid the most recent (and hence lowest) offer that it accepted in the auction.

For all its complexity under the hood, the FCC Incentive Auction's reverse auction was extremely simple for the participating broadcasters. The opening offer for a license was known in advance, and each subsequent offer was automatically 95% of the previous one. As long as the current offer exceeded a broadcaster's value for its license, the obvious move was to accept it (as the broadcaster reserved the right to reject lower offers made later). Once the current offer dropped below the license's value, the obvious response was to reject the offer and go back on the air (as any subsequent offers would be even worse).

The efficacy of the feasibility checker (Section 24.3) had a first-order effect on the government's costs. No harm, no foul when the subroutine timed out with an unpackable set of stations—the FCCDescendingClock algorithm proceeded as it would have anyway. But whenever the feasibility checker timed out on a packable set of stations $S \cup \{v\}$, a pile of money—often in the millions of dollars—was left on the table.[21] The auction could have made a lower offer to the station v, but for the failure of its feasibility checker. You can see why the auction's designers wanted to get the subroutine's success rate up to as close to 100% as possible![22]

24.5   The Final Outcome

The FCC Incentive Auction ran for roughly a year, from March 2016 to March 2017. Nearly three thousand television stations were involved, 175 of which elected to go off the air in exchange for a total compensation of roughly ten billion dollars (an average of around fifty million dollars per license, with high variance across different regions

---

[21] Around 50% of the timeouts occurred with a packable set of stations.

[22] Has there ever been a more direct relationship between an algorithm's running time and huge sums of money?

of the country).[23] Roughly one thousand stations had their channels reassigned.

Meanwhile, the 84 MHz of freed spectrum was reorganized into seven pairs of 5 MHz blocks (one block for uploading, one for downloading). Each of the licenses for sale in the FCC Incentive Auction's forward auction corresponded to one of these seven pairs and one of 416 regions in the U.S. (called "partial economic areas"). The revenue from this forward auction? Twenty billion dollars![24] Most of the resulting profit was used to reduce the U.S. deficit.[25]

The FCC Incentive Auction was a smashing success, and it never would have happened without a cutting-edge algorithmic toolbox for tackling NP-hard problems—the very same toolbox that you can now, after persevering to the end of this book, claim as your own.

The Upshot

☆ The FCC Incentive Auction was a complex algorithm that procured 84 MHz of wireless spectrum used for terrestrial television and repurposed it for next-generation wireless networks.

☆ Its reverse auction decided which television sta-

---

[23]You can check out the full list of results at `https://auctiondata.fcc.gov/`.

[24]Good thing the forward auction revenue exceeded the reverse auction procurement costs—did the government just get lucky? This relates to another question: Who decided that 84 MHz was the perfect amount of spectrum to clear?

The actual FCC Incentive Auction had an additional outer loop, which searched downward for the ideal number of channels to clear (one of the reasons why the auction took so long). In its first iteration (called a "stage"), the auction ambitiously attempted to free up twenty-one channels (126 MHz), sufficient to create ten paired licenses per region for sale in the forward auction. (The twenty-one channels were 30–36 and 38–51; as noted in footnote 4, channel 37 was off limits.) This stage failed badly, with procurement costs roughly eighty-six billion dollars and forward auction revenue only around twenty-three billion dollars. The auction proceeded to a second stage with the reduced clearing target of nineteen channels (114 MHz, enough for nine paired licenses per region), resuming the reverse and forward auctions where they left off in the first stage. The auction eventually halted after the fourth stage (clearing fourteen channels, as described in this chapter)—the first one in which its revenue covered its costs.

[25]Deficit reduction was the plan all along—probably one of the main reasons the bill managed to pass Congress (see footnote 5).

tions would go off the air or switch channels, along with their compensation.

☆ Even with only one available channel, determining the most valuable non-interfering stations to keep on the air boils down to the (NP-hard) weighted independent set problem.

☆ With multiple available channels, merely checking whether a set of stations can all remain on the air without interference boils down to the (NP-hard) graph coloring problem.

☆ On representative instances, a carefully tuned greedy heuristic algorithm reliably returned solutions with near-optimal total value.

☆ Each iteration of this greedy algorithm invoked a feasibility-checking subroutine to check if there was room on the airwaves for the current station.

☆ A descending clock auction was used to implement this algorithm, with offers of compensation falling and stations dropping out over time.

☆ Using presolvers, preprocessing, and a portfolio of eight state-of-the-art SAT solvers, over 99% of the feasibility-checking instances in the FCC Incentive Auction were solved in under a minute.

☆ The FCC Incentive Auction ran for a year, removed 175 stations from the airwaves, and cleared almost ten billion U.S. dollars in profit.

Test Your Understanding

Problem 24.1 *(S)* Which of the algorithmic tools described in Chapters 20 and 21 played no role in the FCC Incentive Auction?

a) Greedy heuristic algorithms

   b) Local search

   c) Dynamic programming

   d) MIP and SAT solvers

Problem 24.2 *(S)* In each round of the FCCDescendingClock algorithm in Section 24.4.3, the stations still in limbo are processed in an arbitrary order. Is the set S of stations returned by the algorithm independent of the order used in each round? (Choose whichever statements are true.)

   a) Yes, provided no two station values w_v are the same and all parameters β_v are set to 1.

   b) Yes, provided all parameters β_v are set to 1 and ϵ is sufficiently small.

   c) Yes, provided no two station values w_v are the same, all parameters β_v are set to 1, and ϵ is sufficiently small.

   d) Yes, provided no two ratios w_v/β_v are the same and ϵ is sufficiently small.

Problem 24.3 *(S)* Before making a lower offer to a station v, the FCCDescendingClock algorithm checks if $S \cup \{v\}$ is a packable set of stations, where S denotes the already-on-air stations. Suppose we reversed the order of these two steps:

```
offer compensation β_v · p to v
if offer refused then          // because p < w_v/β_v
    if S ∪ {v} packable then         // room for v
        S := S ∪ {v}        // v goes back on the air
    else                             // no room for v
        X := X ∪ {v}    // v must stay off the air
```

Suppose we offer compensation to the departing broadcasters as on page 231, with each paid according to the last offer they accepted (the penultimate offer they were given). Is it still true that a broadcaster should accept every offer above its value and reject the first offer below its value?

Challenge Problems

Problem 24.4 *(H)* This problem investigates the solution quality achieved by the WISBasicGreedy and WISGeneralGreedy heuristic algorithms in Section 24.2.3 for the special case of the weighted independent set problem in which the degree $\deg(v)$ of every vertex v is at most Δ (where Δ is a nonnegative integer, such as 3 or 4).

(a) Prove that the independent set returned by the WISBasicGreedy algorithm always has total weight at least $1/(\Delta + 1)$ times the total weight of all the vertices in the input graph.

(b) Prove that the same guarantee holds for the WISGeneralGreedy algorithm with β_v set to $1 + \deg(v)$ for each vertex $v \in V$.

(c) Show by examples that, for every nonnegative integer Δ, the statements in (a) and (b) become false if $1/(\Delta + 1)$ is replaced by any larger number.

Programming Problems

Problem 24.5 Try out one or more SAT solvers on a collection of graph coloring instances, using the formulation (24.1)–(24.3). (Experiment both with and without the constraints in (24.3).) For example, you could investigate random graphs, where each edge is present independently with some probability $p \in (0, 1)$. Or, even better, derive a graph from the actual interference constraints used in the FCC Incentive Auction.[26] How large an input size can the solver reliably process in under a minute, or under an hour? How much does the answer vary with the solver?

---

[26] Available at https://data.fcc.gov/download/incentive-auctions/Constraint_Files/.

Epilogue: A Field Guide to Algorithm Design

With the *Algorithms Illuminated* series under your belt, you now possess a rich algorithmic toolbox suitable for tackling a wide range of computational problems. So rich, in fact, that you might find the sheer number of algorithms, data structures, and design paradigms daunting. When you're confronted with a new problem, what's the most effective way to put your tools to work? To give you a starting point, I'll tell you the typical recipe I use when I need to understand an unfamiliar computational problem. You should develop your own personalized recipe as you accumulate more algorithmic experience.

1. Can you avoid solving the problem from scratch? Is it a disguised version, variant, or special case of a problem that you already know how to solve? For example, can it be reduced to sorting, graph search, or a shortest-path computation?[27] If so, use the fastest and simplest algorithm sufficient for solving the problem.

2. Can you simplify the problem by preprocessing the input with a for-free primitive, such as sorting or computing connected components?

3. If you must design a new algorithm from scratch, get calibrated by identifying the line in the sand drawn by the "obvious" solution (such as exhaustive search). For the inputs that you care about, is the obvious solution already fast enough?

4. If the obvious solution is inadequate, brainstorm as many natural greedy algorithms as you can and test them on small examples.

---

[27]If you go on to a deeper study of algorithms, you'll learn about more well-solved problems that show up in disguise all the time. A few examples include the fast Fourier transform, the maximum flow and minimum cut problems, bipartite matching, and linear and convex programming.

Most likely, all will fail. But the ways in which they fail will help you better understand the problem.

5. If there's an obvious way to split the input into smaller subproblems, how easy would it be to combine their solutions? If you see how to do it quickly, proceed with the divide-and-conquer paradigm.

6. Try dynamic programming. Can you argue that a solution must be built up from solutions to smaller subproblems in one of a small number of ways? Can you formulate a recurrence to quickly solve a subproblem given solutions to the smaller subproblems?

7. In the happy event that you devise a good algorithm for the problem, can you make it even better through the deft deployment of data structures? Look for significant computations that your algorithm performs over and over again (such as lookups or minimum computations). Remember the principle of parsimony: Choose the simplest data structure that supports all the operations required by your algorithm.

8. Can you make your algorithm simpler or faster using randomization? For example, if your algorithm must choose one object among many, what happens when it chooses randomly?

9. If all the preceding steps end in failure, contemplate the unfortunate but realistic possibility that there is *no* efficient algorithm for your problem. Of the NP-hard problems you know, which one most closely resembles your problem? Can you reduce this NP-hard problem to yours? What about the 3-SAT problem? Or any of the other problems in the Garey and Johnson book (page 156)?

10. Decide whether you'd rather compromise on correctness or on speed. If you prefer to retain guaranteed speed and compromise on correctness, iterate over the algorithm design paradigms again, this time looking for opportunities for fast heuristic algorithms. The greedy algorithm design paradigm stands out as the most frequently useful one for this purpose.

11. Consider also the local search paradigm, both for approximately solving the problem from scratch and for a no-downside post-processing step to tack on to some other heuristic algorithm.

12. If you'd rather insist on guaranteed correctness while compromising on speed, return to the dynamic programming paradigm and seek out better-than-exhaustive-search (but presumably still exponential-time) exact algorithms.

13. If dynamic programming doesn't apply or your dynamic programming algorithms are too slow, cross your fingers and experiment with semi-reliable magic boxes. For an optimization problem, try formulating it as a mixed integer program and throwing a MIP solver at it. For a feasibility-checking problem, start instead with a satisfiability formulation and throw a SAT solver at it.

Hints and Solutions

Solution to Problem 19.1: (b),(c). The dynamic programming algorithm for the (NP-hard) knapsack problem is a good example of why (d) is incorrect.

Solution to Problem 19.2: (c). Footnote 2 shows why (a) is incorrect. The spanning trees of a graph can all have distinct total costs (for example, if the edge costs are distinct powers of 2), so (b) is also incorrect. The logic in (d) is flawed, as the MST problem is computationally tractable even in graphs with an exponential number of spanning trees.

Solution to Problem 19.3: (b),(d). Answer (c) is incorrect because polynomial-time solvability is related to but not the same thing as solvability in practice. (Imagine an algorithm with running time $O(n^{100})$ on size-n inputs, for example.)

Solution to Problem 19.4: (a). For example, the dynamic programming algorithm for the knapsack problem shows that (c) and (d) are incorrect.

Solution to Problem 19.5: (e). For (a) and (b), the reduction goes in the wrong direction. Answer (c) is incorrect because some problems (like the halting problem mentioned in footnote 18) are strictly harder than others (like the MST problem). Answer (d) is incorrect when, for example, A and B are the single-source and all-pairs shortest path problems. The formal proof for (e) resembles the solution to Quiz 19.3.

Solution to Problem 19.6: (a),(b),(d). In (a), you can assume without loss of generality that the knapsack capacity C is at most n^6 (why?). For (b), refer to Problem 20.11. For (c), the problem is NP-hard even when the input comprises only positive integers (page 19).

Hint for Problem 19.7: To use a subroutine for the TSP to solve an instance of the TSPP, add one additional vertex, connected by a zero-cost edge to each of the original vertices. To use a subroutine for the TSPP to solve an instance of the TSP, first split an arbitrary vertex v into two copies v' and v'' (each inheriting edge costs from v, and with $c_{v'v''} = +\infty$). Then add two new vertices x, y that are each connected to all other vertices by infinite-cost edges, with the exceptions that $c_{xv'} = c_{yv''} = 0$.

Hint for Problem 19.8: Visit the vertices of G in the same order that depth-first search (from an arbitrary starting vertex) would visit the vertices of T. Prove that the total cost of the resulting tour is $2 \sum_{e \in F} a_e$, and that no tour can have a smaller total cost.

Solution to Problem 20.1: (b). To falsify (a), consider ten machines, ten jobs with length 1, ninety jobs with length 6/5, and one job with length 2. To prove (b), use the assumptions to show that the maximum job length is at most 20% of the average machine load, and plug this into (20.3).

Solution to Problem 20.2: (b). To falsify (a), use a sixteen-element variant of the example in Quiz 20.5; the optimal solution should use two subsets the greedy algorithm five (with worst-case tie-breaking). For (b), the first k iterations of the greedy algorithm match those of the GreedyCoverage algorithm with a budget of k. The approximate correctness guarantee for the latter algorithm (Theorem 20.7) implies that this first batch of k iterations covers at least a $1 - \frac{1}{e}$ fraction of the elements of U. The next batch of k iterations covers at least a $1 - \frac{1}{e}$ fraction of the elements that weren't covered in the first batch (why?). After t batches of k iterations each, the number of still-uncovered elements is at most $(\frac{1}{e})^t \cdot |U|$. This number is less than 1 once $t > \ln |U|$, so the algorithm completes within $O(k \log |U|)$ iterations.

Solution to Problem 20.3: (c),(e),(f). To falsify (a) and (d), take $C = 100$ and consider ten items with value 2 and size 10, together with one hundred items with value 1 and size 1. To falsify (b), consider one item with size and value equal to 100 and a second item with value 20 and size 10. To prove (c), imagine allowing the second greedy algorithm to cheat and fill up the knapsack completely using a fraction of one additional item (with value earned on a pro rata basis). Use an exchange argument to prove that the total value of this cheating solution is at least that of any feasible solution. Argue that the combined value of the solutions returned by the first two greedy algorithms is at least that of the cheating solution (and hence the better of the two is at least 50% as good). To prove (e) and (f), argue that the second greedy algorithm misses out only on the worst 10% (in terms of value-to-size ratio) of the cheating solution.

Solution to Problem 20.4: (a). Each iteration of the algorithm's main while loop chooses one edge of the input graph; let M denote the set of chosen edges. The subset S returned by the algorithm then contains $2|M|$ vertices. No two edges of M share an endpoint (why?), so every feasible solution must include at least $|M|$ vertices (one endpoint per edge of M).

Solution to Problem 20.5: (c). A local search algorithm eventually halts at a *locally* optimal solution.

Hint for Problem 20.6: Store in a heap one object per machine, with keys equal to the machines' current loads. Each machine load update boils down to an EXTRACTMIN operation followed by an INSERT operation with the updated key value.

Hint for Problem 20.7: For (a), you can ignore any jobs after job j (why?). Prove that, if $\ell_j > M^*/3$, each machine is assigned one or two of the first j jobs, with the longest jobs on their own machines and the rest paired up optimally on the remaining machines. For (b), use (20.3).

Hint for Problem 20.8: For (a), use a $k^{k-1} \times k^{k-1}$ grid of elements and $2k - 1$ subsets. For (b), replace each element with a group of N copies of it (each belonging to the same subsets as before). Eliminate ties by adding one additional copy to some of the groups. The choice of N should depend on ϵ.

Hint for Problem 20.9: For example, given an instance of the maximum coverage problem with budget k, ground set $U = \{1, 2, 3, 4\}$, and subsets $A_1 = \{1, 2\}$, $A_2 = \{3, 4\}$, and $A_3 = \{2, 4\}$, encode it using the directed graph

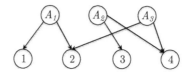

along with the activation probability $p = 1$ and the same budget k.

Hint for Problem 20.10: For (a), verify the properties directly for coverage functions and then use Lemma 20.10. For (b), the primary ingredient is a general version of Lemmas 20.8 and 20.11, and specifically the inequalities (20.7) and (20.15). Let S^* denote an optimal solution and S_{j-1} the first $j-1$ objects chosen by the greedy algorithm. One way to view the right-hand sides of these inequalities is as the sum of the successive marginal values of the objects in $S^* - S_{j-1}$, when added to S_{j-1} one by one in an arbitrary order. The left-hand sides express the sum of the marginal values of the objects in $S^* - S_{j-1}$ when each is added to S_{j-1} in isolation. Submodularity implies that each term in the former sum is at most that in the latter. Where do nonnegativity and monotonicity show up in the proof?

Hint for Problem 20.11: For (a), each subproblem computes, for some $i \in \{0, 1, 2, \ldots, n\}$ and $x \in \{0, 1, 2, \ldots, n \cdot v_{max}\}$, the minimum total size of a subset of the first i items that has a total value of at least x (or $+\infty$, if no such subset exists). For the full solution, see the bonus videos at www.algorithmsilluminated.org.

Hint for Problem 20.12: For (a), every tour can be viewed as a Hamiltonian path (which, as a spanning tree, has total cost at least that of an MST) together with one additional edge (which, by assumption, has a nonnegative cost). For (b), use the triangle inequality to argue that all edge costs in the constructed tree TSP instance are at least as large as in the given metric TSP instance. Using the solution to Problem 19.8, conclude that the total cost of the computed tour is at most twice that of the MST T.

Hint for Problem 20.13: For example, represent the graph using an adjacency matrix (with entries encoding edges' costs) and the current tour using a doubly-linked list.

Solution to Problem 20.14: For (a), the objective function value is always an integer between 0 and $|E|$, and it increases by at least 1 in each iteration. For (b), consider a local maximum. For each vertex $v \in S_i$ and group S_j with $j \neq i$, the number of edges between v and the vertices of S_i is at most that between v and the vertices of S_j (why?). Adding up these $|V| \cdot (k-1)$ inequalities and rearranging completes the argument.

Solution to Problem 21.1: (c).

Solution to Problem 21.2: With columns indexed by vertices of $V - \{a\}$ and rows indexed by subsets S that contain a and at least one other vertex:

$\{a,b\}$	1	N/A	N/A	N/A
$\{a,c\}$	N/A	4	N/A	N/A
$\{a,d\}$	N/A	N/A	5	N/A
$\{a,e\}$	N/A	N/A	N/A	10
$\{a,b,c\}$	6	3	N/A	N/A
$\{a,b,d\}$	11	N/A	7	N/A
$\{a,b,e\}$	13	N/A	N/A	4
$\{a,c,d\}$	N/A	12	11	N/A
$\{a,c,e\}$	N/A	18	N/A	12
$\{a,d,e\}$	N/A	N/A	19	14
$\{a,b,c,d\}$	14	13	10	N/A
$\{a,b,c,e\}$	15	12	N/A	9
$\{a,b,d,e\}$	17	N/A	13	14
$\{a,c,d,e\}$	N/A	22	21	20
$\{a,b,c,d,e\}$	23	19	18	17
	b	c	d	e

Solution to Problem 21.3: (b). Appending an edge (w,v) to a minimum-cost $(i-1)$-hop path P from 1 to w creates a cycle if P already visits v.

Solution to Problem 21.4: (a),(b),(c),(d),(e).

Solution to Problem 21.5: With columns indexed by vertices and rows indexed by non-empty subsets of colors:

$\{R\}$	0	0	$+\infty$	$+\infty$	$+\infty$	$+\infty$	$+\infty$	$+\infty$
$\{G\}$	$+\infty$	$+\infty$	0	0	$+\infty$	$+\infty$	$+\infty$	$+\infty$
$\{B\}$	$+\infty$	$+\infty$	$+\infty$	$+\infty$	0	0	$+\infty$	$+\infty$
$\{Y\}$	$+\infty$	$+\infty$	$+\infty$	$+\infty$	$+\infty$	$+\infty$	0	0
$\{R,G\}$	1	4	1	4	$+\infty$	$+\infty$	$+\infty$	$+\infty$
$\{R,B\}$	2	6	$+\infty$	$+\infty$	6	2	$+\infty$	$+\infty$
$\{R,Y\}$	$+\infty$	$+\infty$	$+\infty$	$+\infty$	$+\infty$	$+\infty$	$+\infty$	$+\infty$
$\{G,B\}$	$+\infty$	$+\infty$	7	3	7	3	$+\infty$	$+\infty$
$\{G,Y\}$	$+\infty$	$+\infty$	8	5	$+\infty$	$+\infty$	5	8
$\{B,Y\}$	$+\infty$	$+\infty$	$+\infty$	$+\infty$	9	10	9	10
$\{R,G,B\}$	5	7	3	5	8	3	$+\infty$	$+\infty$
$\{R,G,Y\}$	9	9	$+\infty$	$+\infty$	$+\infty$	$+\infty$	9	9
$\{R,B,Y\}$	12	15	$+\infty$	$+\infty$	$+\infty$	$+\infty$	15	12
$\{G,B,Y\}$	$+\infty$	$+\infty$	16	13	14	8	8	13
$\{R,G,B,Y\}$	10	17	13	19	15	11	10	11
	a	b	c	d	e	f	g	h

Hint for Problem 21.6: Any vertex j that achieves the minimum in (21.6) appears last on some optimal tour. A vertex k that achieves the minimum in (21.4) immediately precedes j on some such tour. The rest of the tour can be similarly reconstructed in reverse order. To achieve a linear running time, modify the `BellmanHeldKarp` algorithm so that it caches for each subproblem a vertex that achieves the minimum in the recurrence (21.5) used to compute the subproblem solution.

Hint for Problem 21.7: Modify the `PanchromaticPath` algorithm so that it caches for each subproblem an edge (w, v) that achieves the minimum in the recurrence (21.7) used to compute the subproblem solution. Also, cache a vertex achieving the minimum in the last line of the pseudocode.

Hint for Problem 21.8: Throw out the solutions to the size-s subproblems after computing all the solutions to the size-$(s + 1)$ subproblems. Use Stirling's approximation (21.1) to estimate $\binom{n}{n/2}$.

Solution to Problem 21.9: (a) With x_v indicating whether vertex v is included in the solution:

$$\text{maximize} \quad \sum_{v \in V} w_v x_v$$
$$\text{subject to} \quad x_u + x_v \le 1 \quad \text{[for every edge } (u, v) \in E]$$
$$x_v \in \{0, 1\} \quad \text{[for every vertex } v \in V].$$

(b) With x_{ij} indicating whether job j is assigned to machine i, and with M denoting the corresponding schedule's makespan:[28]

$$\begin{aligned}
\text{minimize} \quad & M \\
\text{subject to} \quad & \textstyle\sum_{j=1}^{n} \ell_j x_{ij} \leq M \quad \text{[for every machine } i\text{]} \\
& \textstyle\sum_{i=1}^{m} x_{ij} = 1 \quad \text{[for every job } j\text{]} \\
& x_{ij} \in \{0,1\} \quad \text{[for every machine } i \text{ and job } j\text{]} \\
& M \in \mathbb{R}.
\end{aligned}$$

(c) With x_i indicating whether subset A_i is included in the solution, and y_e whether element e belongs to a chosen subset:[29]

$$\begin{aligned}
\text{maximize} \quad & \textstyle\sum_{e \in U} y_e \\
\text{subject to} \quad & y_e \leq \textstyle\sum_{i : e \in A_i} x_i \quad \text{[for every element } e \in U\text{]} \\
& \textstyle\sum_{i=1}^{m} x_i = k \\
& x_i, y_e \in \{0,1\} \quad \text{[for every subset } A_i \text{ and element } e\text{]}.
\end{aligned}$$

Hint for Problem 21.10: For (a), orient the tour in one direction and set x_{ij} to 1 if j is the immediate successor of i and to 0 otherwise. For (b), show that a union of two (or more) disjoint directed cycles that together visit all vertices also translates to a feasible solution of the MIP. For (c), if edge (i,j) is the ℓth hop of the tour (starting from vertex 1), set $y_{ij} = n - \ell$. For (d), argue that every feasible solution is of the form constructed in (c).

Hint for Problem 21.11: For example, encode the constraint $x_1 \vee \neg x_2 \vee x_3$ as $y_1 + (1 - y_2) + y_3 \geq 1$, where the y_i's are 0-1 decision variables. (Use a placeholder objective function, like the constant 0.)

Hint for Problem 21.12: First preprocess the 2-SAT instance so that every constraint has exactly two literals. (One hack is to replace

---

[28] If the constraints with decision variables on both sides bother you, rewrite them as $\sum_{j=1}^{n} \ell_j x_{ij} - M \leq 0$ for every machine i. These constraints force M to be at least as large as the maximum machine load; in any optimal solution to the MIP, equality must hold (why?).

[29] The first set of constraints force $y_e = 0$ whenever none of the subsets that contain e are chosen. (And if such a subset is chosen, y_e will equal 1 in every optimal solution.)

a constraint like x_i with two constraints, $x_i \lor z$ and $x_i \lor \neg z$, where z is a newly added decision variable. The more practical solution is to iteratively eliminate single-literal constraints—such a constraint forces a variable assignment, which can then be propagated to any other constraints involving that variable.) When only two-literal constraints remain, the key trick is to compute the strongly connected components of an appropriate directed graph (which can be done in linear time; see Chapter 8 of *Part 2*). You're on the right track if your graph has $2n$ vertices (one per literal) and $2m$ directed edges; the given 2-SAT instance will be feasible if and only if every literal resides in a different component than its opposite.

Hint for Problem 21.13: For (b), recall (21.10). For (c), use that ta^* satisfies the constraint while ta does not. For (d), use that a truth assignment and its opposite are equally likely. For (f), use that a running time bound of the form $O((\sqrt{3})^n n^d \ln \frac{1}{\delta})$ for a constant d is also $O((1.74)^n \ln \frac{1}{\delta})$ (because any exponential function grows faster than any polynomial function).

Solution to Problem 22.1: (d). The undirected Hamiltonian path problem reduces to each of the problems in (a)–(c). The problem in (d) can be solved in polynomial time using a variation of the Bellman-Ford shortest-path algorithm (see Chapter 18 of *Part 3*).

Solution to Problem 22.2: (a),(b). The problems in (a) and (b) both reduce to the all-pairs shortest path problem with no negative cycles (for (b), after multiplying all edge lengths by -1), which can be solved in polynomial time using the Floyd-Warshall algorithm (see Chapter 18 of *Part 3*). The directed Hamiltonian path problem reduces to the problem in (c), proving that the latter (and the more general problem in (d)) is NP-hard.

Solution to Problem 22.3: (a),(b),(c),(d). For (b), if the assumed subroutine for the decision version says "no," report "no solution." If it says "yes," use the subroutine repeatedly to delete outgoing edges of s, never deleting an edge that would flip its answer to "no"; eventually, only one outgoing edge (s, v) will remain. Repeat the process from v.

For (d), perform binary search over the target total cost C. The running time is polynomial in the number of vertices and the number of digits required to represent the edge costs, which is polynomial in the input size; see also the discussion on page 19.

Hint for Problem 22.4: Toggle which edges are present or absent.

Hint for Problem 22.5: A subset S of vertices is a vertex cover if and only if its complement $V - S$ is an independent set.

Hint for Problem 22.6: Use one subset per vertex, containing its incident edges.

Hint for Problem 22.7: Use t as the knapsack capacity. Use a_i as both the value and size of item i.

Solution to Problem 22.8: Invoke a subroutine for the maximum coverage problem with the given set system and successively increasing budgets $k = 1, 2, \ldots, m$. The first time the subroutine returns k subsets that cover all of U, these subsets constitute an optimal solution to the given set cover instance.

Hint for Problem 22.9: To reduce the undirected version to the directed version, replace each undirected edge (v, w) with two directed edges (v, w) and (w, v). For the other direction, perform the following operation on each vertex:

Hint for Problem 22.10: For (a), add one additional number to the input. For (b), use the a_i's as job lengths.

Hint for Problem 22.11: Start with a triangle on vertices called t (for "true"), f (for "false"), and o (for "other"). Add two more vertices v_i, w_i per variable x_i in the given 3-SAT instance, and connect them in a triangle with o. In every 3-coloring, v_i and w_i either have the same colors as t and f, respectively (interpreted as $x_i := $ true) or the same colors as f and t, respectively (interpreted as $x_i := $ false). Implement a disjunction of two literals using a subgraph of the form

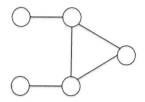

with "inputs" on the left and the "output" on the right. Fuse together two such subgraphs to implement a disjunction of three literals.

Hint for Problem 22.12: Part (b) follows immediately from the reduction in the proof of Theorem 22.7. For part (a), add 1 to each of the edge costs in this reduction.

Solution to Problem 23.1: (a),(b). For (a), for all we know, the TSP (say) can be solved in polynomial time. For (b), for all we know, $\mathcal{P} \neq \mathcal{NP}$ but the Exponential Time Hypothesis is false. For (c), see Problem 23.5. For (d), if any NP-complete problem is polynomial-time solvable, $\mathcal{P} = \mathcal{NP}$ and all such problems are polynomial-time solvable.

Solution to Problem 23.2: Because the TSP is NP-hard (Theorem 22.7), every problem in $\mathcal{NP}$ reduces to it. Hence, if Edmonds's conjecture is false (meaning there is a polynomial-time algorithm for the TSP), so is the P $\neq$ NP conjecture. Conversely, if $\mathcal{P} = \mathcal{NP}$, the search version of the TSP (which belongs to $\mathcal{NP}$) would be polynomial-time solvable. The optimization version of the TSP reduces to the search version by binary search (Problem 22.3) and would then also be polynomial-time solvable, refuting Edmonds's conjecture.

Solution to Problem 23.3: All of them (as you should check).

Hint for Problem 23.4: For (a), compose the reductions. For (c), chain together the two preprocessors and the two postprocessors. To bound the running time, argue as in the solution to Quiz 19.3.

Solution to Problem 23.5: For (a), there is a Levin reduction from 3-SAT to PADDED 3-SAT (add one new variable and the appropriate padding). For (b), check (in linear time) if the input is padded and, if so, use exhaustive search to compute a satisfying assignment or conclude that none exist. Because the size N of a padded instance is at least n^2, this exhaustive search runs in $2^{O(n)} = 2^{O(\sqrt{N})}$ time.

Hint for Problem 23.6: Borrow the trick from the previous problem, with the amount of padding super-polynomial but subexponential. Show that a polynomial-time algorithm for the padded problem would refute the Exponential Time Hypothesis. Using that the padded problem can be solved in subexponential time (why?), prove that a (Cook) reduction from the 3-SAT problem to the padded problem would also refute the Exponential Time Hypothesis.

Hint for Problem 23.7: For part (a), run breadth-first search n times, once for each choice of the starting vertex. For part (b), divide the n variables of a given k-SAT instance into two groups of size $n/2$ each. Introduce one vertex for each of the $2^{n/2}$ possible truth assignments to the variables in the first group, and likewise for the second group. Call the two sets of $2^{n/2}$ vertices A and B. Introduce one vertex for each of the m constraints, along with two additional vertices s and t; call this set of $m+2$ vertices C and define $V = A \cup B \cup C$. (Question: How big can m be, as a function of n and k?) Include edges between each pair of vertices in C, between s and each vertex of A, and between t and each vertex of B. Complete the edge set E by connecting a vertex v of A or B to a vertex w corresponding to a constraint if and only if none of the $n/2$ variable assignments encoded by v satisfy the constraint corresponding to w. Prove that the diameter of $G = (V, E)$ is either 3 or 2, depending on whether the given k-SAT instance is satisfiable or unsatisfiable.

Solution to Problem 24.1: (c).

Solution to Problem 24.2: (c),(d). If in each round of the FCCDescendingClock algorithm there is at most one still-in-limbo station v that would refuse that round's offer (because for the first time, w_v exceeds $\beta_v \cdot p$), the order doesn't affect which stations remain on the air (why?).[30] When the station ratios w_v / β_v are distinct, this condition can be enforced by taking ϵ sufficiently small; hence, answers (c) and (d) are correct. If two stations are poised to drop out in the same round—because of ties between station ratios w_v / β_v or because ϵ isn't small enough—different orderings generally lead to different outputs (as you should check); hence, answers (a) and (b) are incorrect.

---

[30] Though even in this case, the compensation paid to a station going off the air might depend on the order of processing (why?).

Solution to Problem 24.3: Not necessarily, as a broadcaster can in some cases game the system by rejecting an offer higher than its value and receiving more compensation than it would have otherwise. (For example, if the owner of an in-limbo station v learns that the set $S \cup \{v\}$ has become unpackable, the owner should always reject the next offer received.)

Hint for Problem 24.4: For (a), whenever the algorithm includes v in its solution-so-far S, it knocks out from further consideration at most Δ other vertices, each with weight at most w_v. Thus $\sum_{v \notin S} w_v \leq \sum_{v \in S} \Delta \cdot w_v$, which implies the stated bound. For (b), for $v \in S$, let $X(v)$ denote the vertices knocked out from further consideration by v's inclusion in S—that is, $u \in X(v)$ if v is the first neighbor of u added to S, or if u is v itself. By the algorithm's greedy criterion, whenever it includes v in S, $w_v \geq \sum_{u \in X(v)} w_u / (\deg(u) + 1)$. Because every vertex $u \in V$ belongs to the set $X(v)$ for exactly one vertex $v \in S$ (why?),

$$\sum_{v \in S} w_v \geq \sum_{v \in S} \sum_{u \in X(v)} \frac{w_u}{\deg(u) + 1} = \sum_{u \in V} \frac{w_u}{\deg(u) + 1} \geq \frac{\sum_{u \in V} w_u}{\Delta + 1}.$$

Index

Books by Tim Roughgarden

Introductory

Algorithms Illuminated, Part 1: The Basics
(Soundlikeyourself Publishing, 2017)

Algorithms Illuminated, Part 2: Graph Algorithms and Data Structures
(Soundlikeyourself Publishing, 2018)

Algorithms Illuminated, Part 3: Greedy Algorithms and Dynamic Programming (Soundlikeyourself Publishing, 2019)

Algorithms Illuminated, Part 4: Algorithms for NP-Hard Problems
(Soundlikeyourself Publishing, 2020)

Intermediate

Twenty Lectures on Algorithmic Game Theory
(Cambridge University Press, 2016)

Communication Complexity (for Algorithm Designers)
(NOW Publishers, 2016)

Advanced

Selfish Routing and the Price of Anarchy
(MIT Press, 2005)

Algorithmic Game Theory (ed.)
(co-editors: Noam Nisan, Éva Tardos, and Vijay V. Vazirani)
(Cambridge University Press, 2007)

Complexity Theory, Game Theory, and Economics: The Barbados Lectures (NOW Publishers, 2020)

Beyond the Worst-Case Analysis of Algorithms (ed.)
(Cambridge University Press, 2020)